T0275829

Mixed-Species Groups of Animals

Mixed-Species Groups of Animals

Behavior, Community Structure, and Conservation

Eben Goodale
College of Forestry, Guangxi University, Guangxi, China

Guy Beauchamp
Independent Researcher, Montréal, Canada

Graeme D. Ruxton
School of Biology, University of St Andrews, St Andrews, United Kingdom

ACADEMIC PRESS

An imprint of Elsevier

Academic Press is an imprint of Elsevier
125 London Wall, London EC2Y 5AS, United Kingdom
525 B Street, Suite 1800, San Diego, CA 92101-4495, United States
50 Hampshire Street, 5th Floor, Cambridge, MA 02139, United States
The Boulevard, Langford Lane, Kidlington, Oxford OX5 1GB, United Kingdom

Copyright © 2017 Elsevier Inc. All rights reserved.

No part of this publication may be reproduced or transmitted in any form or by any means, elec-
tronic or mechanical, including photocopying, recording, or any information storage and retrieval
system, without permission in writing from the publisher. Details on how to seek permission,
further information about the Publisher's permissions policies and our arrangements with organi-
zations such as the Copyright Clearance Center and the Copyright Licensing Agency, can be found
at our website: www.elsevier.com/permissions.

This book and the individual contributions contained in it are protected under copyright by the
Publisher (other than as may be noted herein).

Notices
Knowledge and best practice in this field are constantly changing. As new research and experience
broaden our understanding, changes in research methods, professional practices, or medical treatment may become necessary.

Practitioners and researchers must always rely on their own experience and knowledge in evaluating and using any information, methods, compounds, or experiments described herein. In using
such information or methods they should be mindful of their own safety and the safety of others,
including parties for whom they have a professional responsibility.

To the fullest extent of the law, neither the Publisher nor the authors, contributors, or editors,
assume any liability for any injury and/or damage to persons or property as a matter of products
liability, negligence or otherwise, or from any use or operation of any methods, products, instruc-
tions, or ideas contained in the material herein.

Library of Congress Cataloging-in-Publication Data
A catalog record for this book is available from the Library of Congress

British Library Cataloguing-in-Publication Data
A catalogue record for this book is available from the British Library

ISBN: 978-0-12-805355-3

For information on all Academic Press publications visit our website at
https://www.elsevier.com/books-and-journals

Working together
to grow libraries in
developing countries

www.elsevier.com • www.bookaid.org

Publisher: Sara Tenney
Acquisition Editor: Kristi Gomez
Editorial Project Manager: Pat Gonzalez
Production Project Manager: Lucía Pérez
Designer: Maria Inês Cruz

Typeset by TNQ Books and Journals

Contents

9. Conclusions

Preface

Mid to late August in northeastern North America means bird migration has started, and avid birders get up each morning with the hopes of seeing something new. A mixed-species flock led by chickadees of the tit family is always a good place to start. Apart from the chickadees themselves, there are the usual suspects: nuthatches or woodpeckers, perhaps some kinglets. But unexpected gems also pass by as one watches the flock: a couple of black-throated green warblers, a northern parula! or a blackburnian warbler! Particular images from childhood remain, forever still and crystallized in our minds: we can still see the tree trunk in the hemlock-dominated forest where a white-breasted nuthatch was busily working upside down, making its "ank-ank" call, when then there was a higher pitched "pink-pink" sound from above, and coming around the trunk was a red-breasted nuthatch, looking down inquisitively at the other nuthatch below. Questions remain, too, in the back of our minds, influencing scientific questions two decades later. Was this a random association between species, or did it represent a coherent group? Why were these particular species found together in the flock? What information was shared between those two birds in that split second so many years ago, and more generally, how do such interspecific interactions influence the lives of birds of these species?

We think this is a useful time to collect information about mixed-species animal groups into a book, in part because the extraordinary experience of observing such a group allows us to understand ecological complexity first-hand. As we will argue later in this book, we believe that mixed-species groups can be important to conserve as systems, rather than constructing conservation plans piecewise for the individual participating species. But firstly we concentrate on a simpler story, though one still relevant to conservation: in this age of biodiversity loss, when an urbanizing human population has fewer and fewer experiences with wilderness, mixed-species groups are accessible phenomena for people to appreciate nature. Documentaries have made some examples, such as mixed herds of African ungulates, widely familiar. More importantly, tropical environments are home to mixed-species systems of many taxa, from birds and primates in forests to fishes on coral reefs, and therefore direct experience is within reach of many people. Even in temperate regions, one needs to visit only a local park at the right time of year to see a mixed-species bird flock or explore the shore of a lake to observe a mixed shoal of fish. These experiences can stimulate in any observer, whether student or researcher, many questions about the interconnectedness of species in a community, and what that means when thinking about the conservation of those communities.

The description of mixed-species groups stretches back centuries, and the literature for certain taxa, such as forest birds, is extensive. Yet, except for a handful of review articles, few publications have sought to compare the adaptive costs and benefits of mixed-species grouping across a broad range of taxa, and even fewer have searched for patterns in the characteristics of such groups in different habitats and taxa. Surely, given the frequent and spectacular nature of these interactions, they are deserving of an introductory description in a book. Undoubtedly, we may not fully capture every aspect of these groups because of the still developing nature of the science and our own limitations. But we hope to establish a foundation of what is known about mixed-species groups, introduce students and future researchers to this topic, and encourage them along new paths of discovery.

The organization of this book should allow readers with different backgrounds and objectives to use it in a variety of ways. In Chapter 1, we describe the scope of the book: defining fluid (not highly persistent) interactions between species of the same trophic level as "mixed-species associations," and then focusing further on "mixed-species groups," which are moving entities, including mostly mutualistic interacting species. The diversity of mixed-species associations is then covered in Chapter 2, with a more fine-grained summary of mixed-species groups in Chapter 3. Readers with a more theoretical interest in evolutionary ecology might want to skip ahead to Chapters 4 and 5, which look in depth at the adaptive costs and benefits of mixed-species grouping, starting with foraging and/or social benefits (Chapter 4), and then moving to antipredatory benefits (Chapter 5).

In the next three chapters, we turn to important concepts in community ecology and conservation that have recently begun to be explored in the mixed-species group context. Chapter 6 investigates communication in groups, as one of the best studied examples of behavior in mixed-species associations. As animals often act as information sources for each other, this subject leads to Chapter 7, in which we examine the roles of species in mixed-species groups, and why some species could be considered keystone species. In turn, such keystone species may be used as targets for the conservation of the whole community, and this subject thus leads to the important topic of the implications of mixed-groups for conservation (Chapter 8).

Finally, in Chapter 9, we collect our thoughts on how the process of writing a book-length treatise on mixed-species grouping has influenced our perception of them. Furthermore, we propose ideas about fruitful directions for future research; we hope that new research may quickly require an expansion of this introductory text.

This book is the collaborative effort of three coauthors. We felt that a collaborative effort was needed to do justice to the wide variety of themes found in the mixed-species grouping literature. We hope that what the book may lack in unity of style as a result is more than compensated by the unique perspective each of us brought to this research.

Eben Goodale would like to first thank his coauthors for the very positive experience working together. He gratefully acknowledges the financial support of the National Science Foundation of the United States for his past work on avian mixed-species groups in Sri Lanka, the 1000 Talents Recruitment Program of the People's Republic of China in helping him come to China, and the National Science Foundation of China (grant # 31560119) for continuing studies on mixed-species groups there. James Nieh and Ashley Ward provided important comments on the empirical sections of the book. Eben is thankful to the series of advisors who have encouraged him to study mixed-species phenomena, including late Donald Griffin, Naomi Pierce, Peter Ashton, Sarath Kotagama, Donald Kroodsma, Bruce Byers, James Nieh, Jin Chen, and Kunfang Cao. He is grateful to all the colleagues, collaborators, and students, who have accompanied him to see these systems in the field or who have collaborated in exploring the data. He also greatly appreciates the support of his family, including his mother Robin Horowitz Milgram, step-father Jerome Milgram, father Alfred Goodale, who first inspired his interest in birds, his loving wife and partner in all of life's adventures, Uromi Manage Goodale, and his son David Parakrama Goodale, with all his energy and endless curiosity.

November 2016
Nanning, China; Montréal, Canada; St Andrews, Scotland

Chapter 1

Introduction

1.1 WHAT IS A MIXED-SPECIES GROUP? DEFINING THE SCOPE OF THE BOOK

What if it were possible to tag every organism in a community with a light, using different colored lights for different species, and then watch their movements in real time? Despite the rapid development of technology to track animals—such as radio telemetry, passive integrated transponder tags, and GPS tracking devices—this kind of experiment will clearly never be possible, given the great abundance of microorganisms even at the scale of a few meters. Yet, let us imagine this as a thought experiment: what kinds of patterns would we see?

Firstly, we would see some species tightly bound together. These are animals like parasites that live inside their hosts; think of the myriad species such as fleas, ticks, flukes, and microorganisms that live on or inside the human body. If we are watching over a longer time frame, then these could also be obligate mutualists, such as corals and zooxanthellae, which require each other for life and then are extinguished together. Also, we would see predator–prey relationships, where one species moves to capture another, shutting down the captured animal's light forever. Notice that all of these relationships have a dyadic nature: the host and parasite, the predator and prey, or two species of obligate mutualists (although multiple parasites may live together on one host).

In contrast to these very tight associations, other animal species will assemble more fluidly, coming together and moving collectively in the same direction, and then in some cases separating again, or in other cases staying together for much of their lives but moving separately (i.e., not on or inside each other). These assemblages can be of one species or two, or many species. What are the driving forces behind these kinds of associations? One possibility is that these animals are not interacting but are attracted to a resource or habitat that they all use. Alternatively, these animals could be coming together in response to each other's presence. These kinds of movements could happen, of course, over a wide range of spatial and temporal scales, some involving the formation of groups over seconds and minutes, whereas others could take weeks or months, such as the migration of animals and their settlement into habitats at the end of the migration.

Considering the wide variety of conceivable patterns, what kinds of groupings are we going to discuss in this book? We confine ourselves to animal groups of multiple species but one trophic level (i.e., animals that consume the same kind

Mixed-Species Groups of Animals. http://dx.doi.org/10.1016/B978-0-12-805355-3.00001-4
Copyright © 2017 Elsevier Inc. All rights reserved.
1

of food resources and not each other), and we exclude interactions where animals move together because they live on or inside each other, whether they be parasites and hosts, or mutualists that always stay together. Such associations, often termed "symbioses," are characterized by their persistence; in contrast, most definitions of groups involve the ability to join and leave fairly frequently (Wilson, 1975). This leaves a wide range of what we call "mixed-species associations" (MSAs) within our scope. They include both phenomena in which animals converge about a resource, which we call "aggregations" (Powell, 1985), and situations where the group itself is the stimulus that attracts animals to join it (Table 1.1).

We particularly focus on a subset of MSAs that we refer to as mixed-species groups (MSGs). Many colloquial terms such as herds, flocks, pods, shoals, swarms, and troops are used to characterize groups in different taxa; animals join these groups to associate with the group as a whole or some animals that participate in them, and these groups are mobile. Other terms, including interspecific groups (Rice, 1956), mixed groups (Eaton, 1953), multispecies groups (Wing, 1946), or polyspecific groups (Gartlan and Struhsaker, 1972), have been used in the past to describe groups with several species, but these terms failed to reach a wide use among biologists working with different taxa. We favor MSGs as a more descriptive term that can be applied to all species.

TABLE 1.1 A Glossary of Terms for Mixed-Species Animal Groups

Term	Definition
Aggregation	A gathering of animals around a resource or a specific location.
Conspecific	An individual animal of the same species as the focal animal.
Heterospecific	An individual animal of a different species as the focal animal.
Interaction	A process related to the spatial proximity of individuals, which brings net benefits to the participants and thus generates a selective force for maintenance or enhancement of the spatial proximity.
Mixed-species association	A gathering of individual animals that belong to the same trophic level and is not as persistent as a *symbiosis*, as individuals frequently join or leave.
Mixed-species group	A group of independently moving animals from more than one species found in close proximity, which interact with one another.
Symbiosis	As association between two species that is characterized by the species living persistently in close proximity.

We use a definition of a group recently advocated by Viscido and Shrestha (2015) to articulate our view of what constitutes an MSG. Viscido and Shrestha argue that groups should be defined globally, rather than using specific criteria about the distances between individuals. Specifically, then for this book, an MSG includes individuals that: (1) are in spatial proximity to each other, but move separately, though in the same direction, and (2) interact with each other, with this interaction being more critical for the formation and maintenance of the group than external factors, such as resource patches that structure aggregations. Another characteristic of MSGs is that usually all individuals gain a net benefit from the interaction. However, we do not make this a part of our definition because individuals also experience various costs in such groups, which may add up to an overall cost of the group for some individuals of some species (see below).

Let us look at these criteria in more detail. The independent movement requirement implies that the group has a degree of fluidity, that is, members have the option of joining or leaving groups. The definition excludes the kind of "in-lock-step" movement of parasites on their hosts or the "always-on" nature of obligate mutualists.

The interaction requirement is important in distinguishing MSGs from aggregations that occur around resources or groups wherein animals migrate at the same time as one another from one point to another location without interacting. Although Viscido and Shrestha (2015) do not explicitly define interactions, here we define an interaction between two individuals to be a process related to the spatial proximity of the individuals, which brings net benefits to the individuals and thus generates a selective force for maintenance or enhancement of the spatial proximity. Given that our earlier definition of an MSG indicates that benefits in MSGs can be asymmetrical, we also allow asymmetrical interactions, where the benefits that generate selective forces may apply only to some individuals of some species. This definition of an interaction does not require one individual to directly change the behavior of the other. For example, by occurring close to one another, individuals of different species can experience a reduction in predation risk by sharing the risk of being attacked by a predator (Hamilton, 1971). Furthermore, by participating in a group, an individual can change the behavior of the predator, hence influencing other group members indirectly.

As to whether all individuals benefit, one problem with including benefits in the definition of an MSG is that the benefits may often be unequal between species, occasionally to the point where one species may not benefit at all or be disadvantaged by the grouping. As evolutionary biologists, we want to explain why grouping behavior evolves, and if such behavior was not beneficial to the animals, we assume, to some degree, that the animals would leave or not join MSGs. Yet, there are some examples where some members could be trapped in an MSG to their detriment. For example, some species that lead MSGs may accrue some costs from the animals that follow them (Hino, 1998; Wrege et al.,

2005; Sridhar et al., 2009), yet if that cost is less than the cost of driving the followers off, or of not being in a group at all, they may remain in MSGs. The benefits to grouping may also vary over time and space, changing from mutualism to commensalism or even parasitism with changing environmental conditions. In addition, in this age of human-induced environmental changes, it is possible that some groups may be maladaptive and evolutionary products of environmental conditions no longer in place. Given these potential cases where not all individuals in MSGs benefit, we make a place in our definitions of MSGs for exploitative or even mutually disadvantageous situations (Section 4.1). We also consider a variety of levels of benefits in MSGs. In some situations, members in MSGs may have greater fitness than those in similarly sized single species groups, but in other situations this may not be the case (Section 4.1).

Another problem in requiring mutual benefits in a definition of MSGs is that the benefits, or the lack thereof, may be difficult to measure. Discrimination of mutualism versus commensalism or mild parasitism can be difficult to accomplish in the field. Take the example of a group of monkeys that feed in the canopy, disturbing leaves that provide a food source to deer that follow beneath them (Newton, 1989). This might well be a case of commensalism, but it would be difficult to rule out the possibility that the deer could actually provide some information about terrestrial threats to the monkeys. To convincingly demonstrate strict commensalism, we would need to demonstrate empirically that the presence or absence of the deer has no effect on the monkeys, a nontrivial empirical challenge.

In summary, we will cover in this book, a range of MSAs between animals of more than one species of the same trophic level, where the gathering of animals is fluid, with individuals joining and leaving; we will especially emphasize associations where a suite (more than two) of species is involved. The diversity of such MSAs is surveyed in Chapter 2. We will further focus on MSGs, where the individuals are moving separately although in the same direction, in which the group members are interacting with one another, and in which most members benefit from the association, with the diversity of such MSGs explored in Chapter 3.

In this book, we cover all taxa in which animals gather in groups involving more than one species. Plants and other sessile organisms, such as those of the intertidal zone, will not be a focus of the book, because they tend to compete heavily for space (Yodzis, 1978), a requirement quite distinct from most grouping animals, and they almost always live in mixed-species communities. Yet, benefits of associating with other species are quite similar in plants and the animals we consider here. For example, grouping with other plants can change the risk of herbivory by diluting the chance of encounter with an herbivore, and grouping with a plant particularly susceptible to herbivores, or one that is unpalatable, can also change the risk of herbivory up or down, respectively (Section 5.3). Given these similarities in the benefits of association, and that plants actually can influence their neighbors through communication and quasi behavior (Karban, 2008), we include some examples from plants in the chapters on the adaptive benefits to MSAs (Sections 4.5, 4.6 and 5.3).

1.2 HISTORICAL PERSPECTIVE ON RESEARCH ON MIXED-SPECIES GROUPS

Of all the phenomena we cover in this book, forest bird MSGs are the most diverse (species-rich), and this attribute is heightened in the tropics (Section 3.5). Hence, it is fitting that the first published reference to MSGs to our knowledge is by Bates (1863), who evocatively in his "The Naturalist on the River Amazons" wrote "One may pass several days without seeing many birds, but now and then the surrounding bushes and trees appear suddenly to swarm with them. There are scores, probably hundreds, of birds all moving about with the greatest activity… The bustling crowd loses no time, and although moving in concert, each bird is occupied on its own account in searching bark, or leaf, or twig… In a few minutes, the host is gone, and the forest path remains deserted and silent as before."

Indeed, in the 19th and the first half of the 20th century, natural history books, both reports of field expeditions and compendiums on the animals of certain areas, or certain taxa, were one of the major repositories of information on MSGs and other MSAs. Only by the 1960's was there a body of scientific journal articles that concentrated on the MSG phenomenon, and at that time, it was almost exclusively about terrestrial forest bird MSGs. From that time to the present the literature has increased in a roughly linear way (Fig. 1.1); the slowing of that increase in the past 10 years may represent the fact that newer articles have not fully been recognized as yet and referenced by others. Literature on other kinds of MSAs is scattered. Because of the vast breadth of the phenomenon (e.g., ranging from colonial spiders to scavenging vultures), it is difficult to synthesize, and we will thus concentrate on the MSG-specific literature here.

Almost all early articles that described MSGs speculated on their adaptive benefits. Early reviews include that by Rand (1954) on birds, and the review on all taxa by Morse (1977), which lists many possible benefits to such groups. Later reviews of adaptive benefits include Diamond (1981) for all taxa; Harrison and Whitehouse (2011) for forest and waterbirds; Powell (1985), Greenberg (2000), and Colorado (2013) for forest birds; Terborgh (1990) for terrestrial birds and primates; Stensland et al. (2003) for all mammals; Cords and Würsig (2014) for primates and cetaceans; and Lukoschek and McCormick (2000) for fish. By the 2000's the literature on forest bird MSGs was large enough to allow quantitative meta-analyses (Sridhar et al., 2009, 2012; Goodale and Beauchamp, 2010). Because of all this research effort and consistent findings—for example, the importance of antipredatory benefits for birds (Thiollay, 1999; Sridhar et al.,

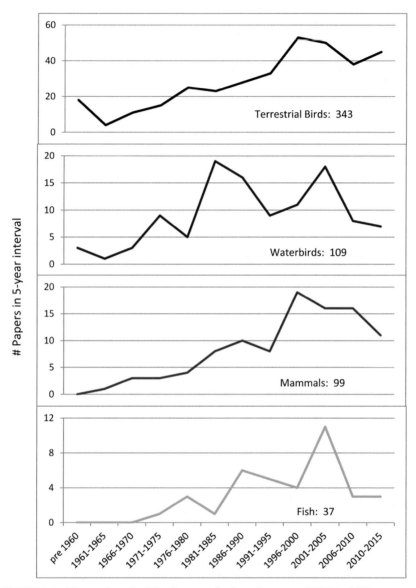

FIGURE 1.1   The number of publications on mixed-species groups (MSGs) of different taxa overtime. To be included in the analysis, the publication had to focus on the mixed-species phenomenon. Note that for waterbirds, some groups may actually be closer to aggregations than to moving MSGs (Section 3.5).

2009)—we see a trend now and in the future for new studies to explore specific questions about MSGs rather than simply speculating on their adaptive benefits.

Some questions about MSGs have been investigated throughout the development of the literature. For example, what makes some species "nuclear," or important to flock formation or maintenance, has attracted attention from early studies (Winterbottom, 1949; Moynihan, 1962) to the present (Smith et al., 2003; Goodale and Kotagama, 2005b; Farley et al., 2008; Srinivasan et al., 2010). Another consistent feature of the literature is questioning whether species associations in MSGs show patterns significantly different from those expected by chance. The question of whether MSGs are simply a result of individuals of different species accidentally meeting was explored in primates by Waser (Waser and Case, 1981; Waser, 1984) and in waterfowl by Silverman et al. (2001). In both of these cases, the percentage of individuals participating in MSGs was low, and MSGs did not pass such tests of significance (i.e., they appeared to be random associations of animals that met by chance at one location at one time), although animals of the same taxa passed such tests and were more clearly associated in other locations (e.g., Holenweg et al., 1996). For dolphins in a vast ocean, or birds in a large forest, often eerily silent before one encounters a mixed flock as in Bates' (1863) description, it seems intuitively clear that MSGs are not a random collection of species, although to our knowledge this has not actually been quantitatively investigated. However, numerous publications have explored a more complex question, whether pairs of species are found together in flocks more than expected by chance, traditionally using statistical tests based on tables of association frequencies (Bell, 1983; Hutto, 1994). Recently, null models and other statistical approaches, such as social networks, have been used to determine whether such associations are expected or not (Graves and Gotelli, 1993), and what rules animals follow in joining MSAs (Farine et al., 2014). Such an approach has proven successful in quantifying which species are most important to MSGs (Srinivasan et al., 2010; Sridhar et al., 2013; Kiffner et al., 2014; see Sections 7.2.2.3 and 7.2.2.4).

In contrast, other topics have only recently expanded in the literature. For example, looking just at the terrestrial bird literature, the percentage of descriptive articles (those that describe bird MSGs in a particular place) is lower than it was initially, and conversely the number of articles that discuss how these systems respond to human-induced disturbances, has increased sharply since 1990 (Fig. 1.2). Traditionally, descriptions of MSGs in birds have been made in mature forests, and even those done in human-dominated landscapes really did not explore the effect of disturbance on the system (e.g., Ulfstrand, 1975). However, within the last few years, there have been a burst of papers looking at MSGs over a gradient of fragment size or a gradient of land-use intensity (see Chapter 8). Another subject that has attracted increasing attention is that of communication in MSGs (see Fig. 1.2 and Chapter 6). In addition, several articles have questioned the traditional characterization of MSGs in birds as interactions between monolithic species and rather demonstrated a high degree of within-species variation in roles (Hino, 2000; Farine and Milburn, 2013).

These recent developments are reflected in a "word cloud" of recent topics in the MSG literature (Fig. 1.3) with the prominence of "individuals," "conspecifics," "networks," and "conservation." The organization of this book also

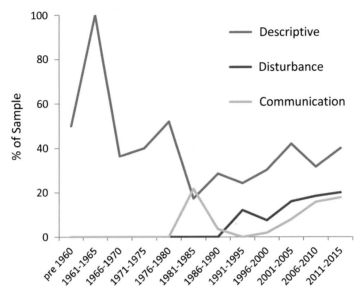

FIGURE 1.2 The prominence of different topics overtime in articles about terrestrial bird mixed-species groups (MSGs). *Descriptive,* included data on the composition of MSGs; *Disturbance,* focused on how MSGs respond to anthropogenic disturbance; *Communication,* focused on some aspect of communication in MSGs. The three categories are not mutually exclusive. Sample size is the same as Fig. 1.1 top panel.

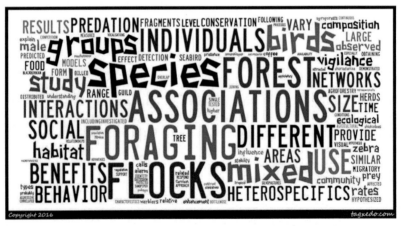

FIGURE 1.3 A "word cloud" of frequent words used in recent publications on mixed-species groups (MSGs). The data used were the abstracts from 47 articles published between 2013 and 2015. Some words specific to MSGs were removed (i.e., associations, species, different, mixed, groups, heterospecific), as well as words with little meaning when isolated (i.e., important, explain, change). Produced by Tagxedo (www.tagxedo.com).

reflects this trajectory of the literature, with the empirical Chapters 2 and 3 first, followed by theoretical explanations of grouping (Chapters 4 and 5), and then by chapters on communication (Chapter 6), leadership (Chapter 7), and conservation (Chapter 8).

Chapter 2

A Diversity of Mixed-Species Associations

2.1 CLASSIFYING MIXED-SPECIES ASSOCIATIONS

In the previous chapter we defined a mixed-species group (MSG) to be a gathering of multiple species of animals in which the participants move separately but in spatial proximity and where the interaction between group members is more important to the formation and maintenance of the group than external factors; we also discussed how most members usually benefit. Yet, at the same time, we acknowledged that there were many other types of associations that do not meet all these requirements but still share characteristics with phenomena that do. In this chapter, we step back and describe this diversity of mixed-species associations (MSAs), looking for common themes. As discussed in the introduction, we will emphasize MSAs that involve a suite of species, rather than symbiotic interactions composed of two species in a coevolved and persistent association, although we will note some similarities between MSAs and symbiotic associations as we go. The MSAs that we cover are not obligate associations, and individuals of the interacting species can flourish without ever participating in MSAs. This focus may bias our material away from invertebrates because invertebrate symbioses are so varied and complex. Our aim is to show the range of interactions that can occur between animals at the same trophic level; however, given the breadth of the material, we are not able to be comprehensive about all the types of interactions between various animal taxa that do occur. In any case, our aim is to draw attention to general trends and principles, rather than offer exhaustive encyclopedic coverage.

We organize this chapter by loosening the different criteria by which we defined MSGs in Chapter 1. We first investigate different kinds of MSA in which the ultimate explanation for the association lies in the presence of external factors and not the interactions between group members. We will see that these aggregations (Section 2.3) can be formed around habitat patches (Section 2.3.1) or resources such as food or water (Section 2.3.2), or the presence or absence of predators (Section 2.3.3). Aggregations that follow moving resources (e.g., fish populations) may appear similar to MSGs. Animals moving on the same migratory pathways, although rarely interacting, can also be found in close proximity (Section 2.3.4). For aggregations,

Mixed-Species Groups of Animals. http://dx.doi.org/10.1016/B978-0-12-805355-3.00002-6
Copyright © 2017 Elsevier Inc. All rights reserved.

although interactions between species do not drive the association, they continue to occur, whether they be in the sharing of information or in competition over extracting the resource, and we explore such interactions because they are reminiscent of those found in moving MSGs. To set the foundation for this idea, we first explore phenomena in which animals interact with one another but are not closely associated in space (Section 2.2).

In Section 2.4, we explore MSAs where the existence of the group depends on species interactions but which are mostly stationary. First, we examine situations in which animals aggregate in space around an absence of predators, similar to the previous section, but now specifically focusing on predator-free space maintained by an aggressive, "protective" heterospecific (Section 2.4.1). Next, we discuss mixed-species colonies and roosts of vertebrates (Section 2.4.2), where the site is stationary, although the animals may continually be moving away from and returning to the site. Animals in such MSAs usually disperse to forage. Given that most moving MSGs are centered on foraging, it seems reasonable to discuss these colonies separately. In Section 2.4.3, we investigate cleaning mutualisms. These interactions can be complex, especially in marine environments, and involve a suite of species. As they are generally stationary, usually taking place at "cleaning stations," we integrate them with other nonmoving MSAs; although if they do move, they are akin to a moving aggregation.

2.2 INTERACTIONS BETWEEN SPECIES WITHOUT ASSOCIATION

Let us start this discussion of the range of MSAs with the absence of association in space. Our motivation here is to recognize that even in such situations animal species of the same trophic level can still behaviorally interact.

One of the major ways in which species interact is by using heterospecifics as sources of information. Cues provided by the presence or behavior of an animal are referred to as "public information" (Danchin et al., 2004; Dall et al., 2005). Public information can be used to find a resource or a habitat type and therefore may lead to aggregations, as will be discussed in the next few sections. However, information sharing of this sort need not lead to aggregations. For example, many species of animals can recognize the alarm calls—calls made upon detection of threat—of other species in their communities (Magrath et al., 2015a). This behavior is often referred to as "eavesdropping" on heterospecific information, because for the most part the animals that make these calls are directing them toward a conspecific audience (Section 6.2). Animals then use this information to choose an appropriate response, such as fleeing or freezing, or becoming more vigilant while performing other behaviors (Beauchamp, 2015; Magrath et al., 2015a).

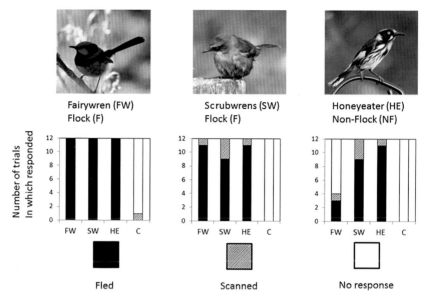

FIGURE 2.1   Group living does not predict response to heterospecific alarm calls. Three species of birds were tested on their responses to each other's vocal alarm calls. Superb fairy-wrens and white-browed scrubwrens participate in nonbreeding MSGs, whereas New Holland honeyeaters do not. However, they all respond to each other's calls. The control (C) was the contact call of a parrot. *Adapted from Magrath, R.D., Pitcher, B.J., Gardner, J.L., 2009. An avian eavesdropping network: alarm signal reliability and heterospecific response. Behavioral Ecology 20, 745–752. Pictures from Wikimedia Commons, by Ron Knight (fairy-wren), benjamint444 (scrubwren), Louise Docker (honeyeater).*

Members of MSGs are well known to respond to each other's alarm calls (e.g., Sullivan, 1984; Goodale and Kotagama, 2005a; Section 6.2.4). But it also seems that animals are capable of responding to the alarm calls of other species even if they are not associated in MSGs but just live in the same general area. For example, Magrath et al. (2009) compared response to heterospecific alarm calls between two avian species that associate together in nonbreeding MSGs and one that does not. The three species all showed some response to each other, and responses were influenced less by grouping behavior than by the relevance of the information about predators to the different species (Fig. 2.1). A recent community-wide study of alarm calling in French Guiana showed that birds react to heterospecific alarm calls of other species, regardless of whether the individuals involved associate in MSGs (Martínez et al., 2016). Indeed, completely unrelated and nonassociating species, such as non-social reptiles or mammals, can respond to the alarm calls of birds that live in the same area (Vitousek et al., 2007; Lea et al., 2008; Fuong et al., 2014). Recently, it has been

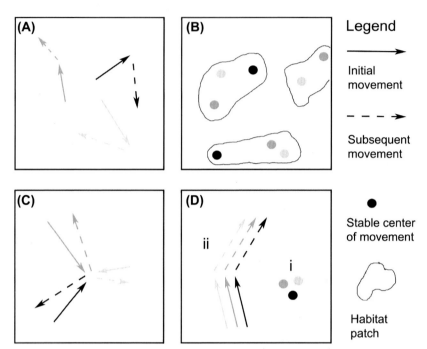

FIGURE 2.2 **The role of interspecific information transfer in the organization of animal communities.** Different shades represent animals of different species; *solid lines* are movement in response to acquisition of information from a heterospecific; and *dashed lines* are subsequent movement back to normal activity. (A) In the case of some information transfer among members of a community, spatial relationships may be unaffected. (B) Heterospecific information can be used by animals to select habitat. (C) Heterospecific information can be used by animals to assemble around a resource or predator. (D) Heterospecific information can be essential for the cohesion of stable colonies (i) or moving MSGs (ii). *Modified from Goodale, E., Beauchamp, G., Magrath, R.D., Nieh, J.C., Ruxton, G.D., 2010. Interspecific information transfer influences animal community structure. Trends in Ecology and Evolution 25, 354–361, with permission.*

shown that bees that use the same flower resources also respond to heterospecific alarm pheromones (Wang et al., 2016). Communities of aquatic organisms can recognize the alarm chemicals of heterospecifics in a somewhat analogous way (Ferrari et al., 2010).

In summary, interspecific information transfer is not solely the domain of MSAs. In a previous review, we summarized the role that interspecific information has on the spatial organization of animal communities (Fig. 2.2; Goodale et al., 2010). We have just described examples where interspecific communication continues in the absence of spatial association (Fig. 2.2A). Interspecific information is also used by animals to find suitable habitats (Fig. 2.2B; Section 2.3.1), to form aggregations (Fig. 2.2C; Sections 2.3.2 and 2.3.3), and can influence the stability of mixed-species colonies (Fig. 2.2D; Section 2.4.2) or moving MSGs (Fig. 2.2D; Chapter 3).

2.3 ASSOCIATION OF SPECIES DESPITE LACK OF INTERACTION

2.3.1 Aggregations in Habitat Patches

One of the most common types of aggregations, at a large scale, is the association of different species within a shared habitat. Obviously, the large topic of habitat selection that explains such associations is beyond the scope of this book. But here we want to point out that although the associations of species in habitats may be the result of individuals of different species choosing the same habitat entirely independently, it can also be the result of animals using heterospecifics as cues of appropriate habitat. Such behavior is known as "heterospecific attraction," analogous to "conspecific attraction," a more well-known phenomenon in which animals tend to prefer habitat where other conspecifics have settled earlier (Stamps, 1988). Heterospecific attraction has been most explored for migrating birds: can they judge appropriate habitat to settle in based on the acoustic information made by other species already settled? Experiments in which songs sung by birds are played back from speakers indicate that migrants can use heterospecific cues to settle, particularly for habitat specialists that use discrete and rare habitat types such as wetlands (Herremans, 1990; Mukhin et al., 2008). Furthermore, experiments in which the abundance of a resident bird is artificially increased in patches show that migrant birds will then chose such patches to breed themselves (Mönkkönen and Forsman, 2002). Heterospecific attraction thus implies that distributions of species can be aligned together or clumped, even in the absence of the clumping of resources.

A demonstration of how detailed information can be transferred among members of an animal community is the work of Seppänen et al. exploring how the nesting preferences of birds are influenced by information derived from heterospecifics. In a series of studies, the researchers set out nest boxes in early spring in Scandinavia that were selected by resident tit species. The researchers then painted a symbol in the front of all selected next boxes—either a triangle or a circle—in a particular patch of forest, and for each box another empty box was placed with the opposite symbol in a nearby tree. The effect of this manipulation for migrant species of flycatchers, which start nesting about 2 weeks after the tits, was that it appeared that all the tits in their forest patch had nested only in a box associated with one symbol. The researchers then presented a pair of new nest boxes for the flycatchers, one with each symbol. In a quite spectacular result, flycatchers strongly preferred nest boxes associated with the symbol they had seen the tits prefer (Seppänen and Forsman, 2007). The researchers then went one step further: they manipulated clutch sizes of the tits so that large clutch sizes were associated with one symbol and small clutch sizes were associated with the other symbol. Their results in this experiment demonstrated that the flycatchers used nests with the symbol associated with large clutch sizes for tits (Loukola et al., 2013; Fig. 2.3).

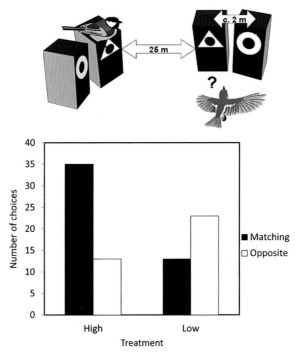

FIGURE 2.3 Animals use information gleaned from heterospecifics to make important decisions. The clutches of resident great tits in Scandinavia, nesting in nest boxes, were manipulated so that they had either high clutch size (13 eggs) or low clutch size (5 eggs). A symbol (*circle* or *triangle*) was then randomly assigned to the nest box and painted on the front of the box; another empty nest box was placed nearby with the other symbol. The migrant pied flycatcher arrives later in the spring and inspects great tit nests; the experimenters offered the flycatchers the choice of two identical nest boxes, one painted with a circle and the other with a triangle. If pied flycatchers were near tits that had high clutch size, they chose the nest box with the same symbol as the tits, but if they were near tits with low clutch size, they chose the opposite symbol. *Adapted from Loukola, O.J., Seppänen, J.-T., Krams, I., Torvinen, S.S., Forsman, J.T., 2013. Observed fitness may affect niche overlap in competing species via selective social information use. American Naturalist 182, 474–483; graphics, copyright Janne-Tuomas Seppänen and Olli Loukola, used with their permission.*

Although this example does not result in the formation of MSAs, it is telling how closely animals inspect the spatial positioning and reproductive success of other species. Seppänen et al.'s work is also instructive in showing that not all species are equal in the quality of information that they provide to other species. At their study site, the year-round resident tits possessed more current information about the area than migrant species, which gradually appeared in the area in the late spring. Later, we will see that species in aggregations can also differ in the type of information they provide, and thus some species can have special roles as "initiators" of such MSAs. The similar subject of what species play important roles in moving MSGs is discussed in detail in Section 7.2.3.

The key difference between animals being found together within habitats and the kinds of aggregations discussed below is that of the scale in space and time. The decision rules the animal uses during habitat selection will influence where it stays for at least one breeding season or potentially for as long as its lifetime. The area it selects will be large relative to the animal—a territory or a home range. In contrast, aggregations can occur over hours or days but generally last no longer than that because the resource becomes exhausted. Depending on the resource, such aggregations can be over very large areas, as much as hundreds of square kilometers in the case of large seabird flocks (Hunt et al., 1988), or small areas, such as a particular fruiting tree within a monkey's home range.

2.3.2 Aggregations Around a Resource

Aggregations around a resource are common across the whole animal kingdom. Large numbers of animals attracted to scattered water holes, as on the African savannah, easily come to mind. Animals respond to any temporary food resource in a similar way, whether it be a mass-fruiting tree, nectar-producing flowers, or carrion that attracts hyenas or vultures; smaller organisms may respond at a different scale, such as guilds of insects that are attracted to a pile of feces. Animals can also gather at a resource that provides a rare nutrient, such as mineral licks for vertebrates (Klaus and Schmidg, 1998; Lee et al., 2010) or salt sources for invertebrates (Otis et al., 2006). In the pelagic environment, animals can become passively clustered due to currents or topographical features (Ritz et al., 2011).

MSAs also occasionally occur at places where individuals of one sex, usually male, display to attract the other sex ("leks," Höglund and Alatalo, 1995). Such MSAs may occur due to the location being particularly good habitat or well trafficked by one sex. Examples of mixed-species leks have been described in both birds (Gibson et al., 2002) and butterflies (Srygley and Penz, 1999).

Although the examples above, and most aggregative behavior, are based on stationary resources, sometimes the resources can move. Aggregations above a moving resource may appear similar to a moving MSG; indeed, the only evidence one can gather to distinguish between these two kinds of groups is to show a tight correlation between the prey movement and the group movement. For example, seabirds aggregate above fish schools in the ocean (Hoffman et al., 1981). Because the fish remain close to the surface for only a short time—Hoffman et al. described aggregations averaging from 100 to 800s in duration—the seabirds do not travel long distances with them, but their groups are constantly forming and breaking apart. Given their similarities with MSGs, we will examine some of these types of associations, particularly those of waterbirds, in Chapter 3. These moving aggregations also have a lot in common with groups that follow "beating" or "driving" animals, phenomena that will be described in Section 3.6.1.

Although animals are not joining aggregations to associate with other animals, they can use the presence of other animals to detect the resource. Such a choice of foraging location based on the presence of other animals is known as "local enhancement" (Thorpe, 1956; Section 4.2.1). For example, certain species of gulls are classified as initiators of MSAs of seabirds because they are the first species to find groups of fish, and other species use them as cues to join the assemblages (Sealey, 1973; Hoffman et al., 1981; Camphuysen and Webb, 1999). Similarly, researchers studying aggregations of large wading birds argue that snowy egrets are especially skilled at detecting fish schools in marshes and are joined by other heron species to make MSAs (Caldwell, 1981; Master, 1992; Smith, 1995). Bees are known to use the presence of conspecifics or heterospecifics to choose or reject flowers (Slaa et al., 2003), and further they can even use chemical cues left inadvertently by other insects on flowers to make such kinds of decisions (Stout and Goulson, 2001).

Dominance hierarchies among the animals attracted to a resource can also play a part in the association's assembly. For example, some larger gulls come later to aggregations of seabirds, particularly those that follow boats, displacing the smaller birds that arrived earlier (Hudson and Furness, 1988; Camphuysen and Webb, 1999). In an invertebrate example, stingless bees leave odor marks on the flowers they visit. Usually this is used by conspecifics, but James Nieh et al. (2004) showed that it can also be used by heterospecifics. In this case, a dominant species of stingless bee followed the odor marks of a subordinate species, and then when it found the resource, usurped it by force. Similar interactions occur in a guild of competing ants (Binz et al., 2014).

The occurrence of dominance interactions between species is another indication that aggregations are not unstructured communities. Rather, interspecific interactions can constrain which species can occur in MSAs, and these interactions can also shape the spatial organization of species around the resource. For example, in a study of aggregations around fruiting trees in Cameroon, body size predicted which species attacked or displaced another in trees, with a dominance hierarchy, from the lowest to the highest, of small birds, squirrels, large birds such as hornbills, and monkeys (Fig 2.4; French and Smith, 2005). This dominance hierarchy affected how much the animals ate, with dominant species eating more fruits per visit. Indeed, interspecific aggression associated with dominance is believed to be so strong among birds at some fruiting trees that subordinate species might deceptively gain access to trees by mimicking the plumage of dominant species (Béland, 1977; Diamond, 1982; Prum, 2014).

2.3.3 Aggregations Influenced by Predation

Aggregations may be shaped as much by predators as by resources. In one way this is the flip side of the same coin: predator-free space can also be viewed as a resource. In some cases, where predators are stationary (e.g., they have nests), animals may place their territories or their own nests accordingly; in Section

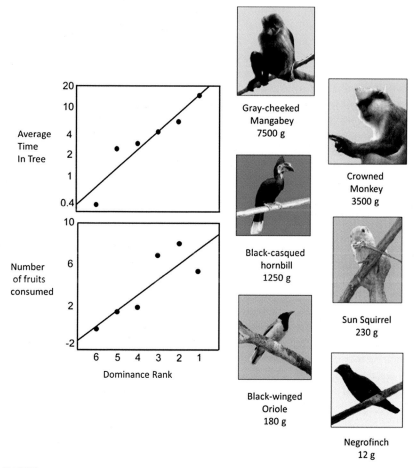

FIGURE 2.4 Interspecific dominance can structure aggregations at fruiting trees. In an African system, the interspecific rank of an animal, correlated with its size, strongly influenced the amount of time it was able to stay in a tree, and the amount of fruit consumed. Some representative species, showing the full range of size of the animals, are shown. *Adapted from French, A.R., Smith, T.B., 2005. Importance of body size in determining dominance hierarchies among diverse tropical frugivores. Biotropica 37, 96–101. Pictures from Wikimedia Commons; mangabey by Marie de Carne, crowned monkey by Trisha Shears, hornbill by Dick Daniels, squirrel by Astrid Spranz, oriole and negrofinch by Francesco Veronesi.*

2.4.1, we will examine animals that place their nests near predators and thereby reduce nest predation (as the predators keep away nest predators). As predators are usually mobile, however, places without predators may not be stationary in the same way that patches of resources are stationary. As described in depth later in Chapter 5, joining moving monospecific groups or MSGs may be an efficient way of finding space where predators are not initially present and can be detected when they approach.

To further discuss predation's influence on the spacing of prey, let us take an example of tadpoles that live in ponds in the Madagascan rainforest. Small ponds there form within riverbeds or on the forest floor and may dry up or become connected with other ponds seasonally. Frogs of six species can be found in these ponds, with up to four species in one pond, including 200–5000 individuals (Glos et al., 2007a). At first, this seems simply an aggregation based on the availability of water, but closer inspection reveals something more complicated: these tadpoles form moving MSGs that occur more consistently in clear ponds without much plant vegetation, where tadpoles are more visible to their predators (Glos et al., 2007a). This suggests that these MSGs are adaptations to cope with high predation pressure; subsequent experiments indeed demonstrated that tadpoles would group in the presence of fish or tadpole homogenate (like many aquatic animals, tadpoles use the chemicals that come from a puncture of the skin, also known as "Schreckstoff," to detect the occurrence of predation; Glos et al., 2007b).

Generally, tadpoles cannot move from pond to pond, so those in clear ponds have little alternative to come together into MSGs. But what happens in those ponds where there is some vegetation, perhaps localized to one part of the pond? Rather than forming groups, all the tadpoles try to stay hidden, close to the vegetation (Fig. 2.5). This clumping around the vegetation is itself a form of an MSA around predator-free space. This example suggests that these kinds of diffuse MSAs in protected predator-free habitats might be difficult to observe and might be more common than expected.

Nesting aggregations are also often believed to be related to predator-free space. Mixed-species colonies of waterbirds may be placed where predators

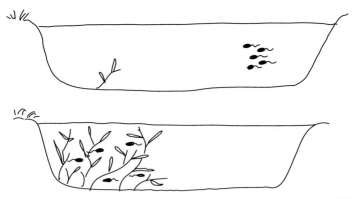

FIGURE 2.5 Aggregations that form around predator-free space may be difficult to detect. In the top panel, tadpoles of different species aggregate in ponds without vegetation. In the bottom panel, the tadpoles are also clustered together in space, but this time they are close to the vegetation so that their association might be not obvious to observers. *Inspired by Glos, J., Dausmann, K.H., Linsenmair, E.K., 2007a. Mixed-species social aggregations in Madagascan tadpoles—determinants and species composition. Journal of Natural History 41, 1965–1977.*

and nest predators have difficulty reaching them, such as offshore islands (Burger, 1981). Nests of other species may be clustered in areas not only where resources are plentiful but also where nest predators are few (Forsman and Martin, 2009). Similar reasons may underlie the placement of multispecies communal roosts (e.g., Burger et al., 1977a for birds; Rainho and Palmeirim, 2013 for bats), although species may come together at such roosts not only because of predator-free space but also because appropriate areas (e.g., large caves) are simply rare (Beauchamp, 1999). For further discussions of colonies and roosts, see Section 2.4.2.

Predators can also attract prey animals around them. This phenomenon is well-known for birds, which gather around and harass ("mob") predators, especially at times when these predators are less dangerous, such as when owls are found during the day (Curio, 1978). The activity often attracts heterospecifics and conspecifics (e.g., Hurd, 1996; Nocera et al., 2008; Suzuki, 2016). Indeed, research by Krams and Krama (2002) suggests that birds can exhibit interspecific reciprocity in mobbing behavior. The researchers compared mobbing activity in sedentary communities, where individuals are familiar with individual heterospecifics, and migratory ones, in which individuals are less familiar with one another, limiting the possibility of reciprocity. Mobbing activity was higher in sedentary communities, but this difference disappeared after the migratory birds became established in the community for 2 weeks (Fig. 2.6).

Some species may be particularly important as initiators of mobbing and hence may play keystone roles in community awareness of predators (Turcotte and Desrochers, 2002; Sieving et al., 2004; Nolen and Lukas, 2009), similar to the initiating species around food resources discussed above. Apart from birds, mobbing behavior is also known in fish (Dominey, 1983) and mammals (as reviewed by Caro, 2005) and may sometimes be used in a mixed-species context in these taxa (e.g., for primates, Gautier-Hion and Tutin, 1988).

2.3.4 Aggregations During Migration

Before we leave aggregations that are based on external factors (such as resources and predators) rather than interactions with the other members of the association, we should consider the possibility that different species of migratory animals may be found together because they are traveling along the same pathways, rather than interacting. Landfalls of migratory birds usually include a variety of species (e.g., Moore et al., 1990). Similar MSAs might be found among aquatic organisms that undergo diel migration, moving up and down in the water column, although it has been suggested that species that socially aggregate in the oceans may not be engaged as much in diel migration because the functions of such migration (which are thought to be the avoidance of high predation rates in daylight close to the surface) and aggregation are similar (Ritz, 1994).

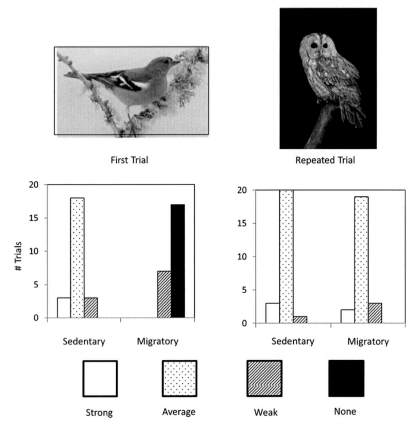

FIGURE 2.6 The level of mobbing activity can be influenced by interspecific reciprocity. Chaffinches were more likely to mob an owl mount if they mobbed as a member of sedentary bird communities, compared to communities composed of migratory individuals, just arrived in an area and therefore not familiar to the chaffinches. When the experiment was conducted 2 weeks after the migrants settled ("repeat" trials), mobbing activity increased. The level of mobbing activity was categorized as strong, average, weak, or none. *Adapted from Krams, I., Krama, T., 2002. Interspecific reciprocity explains mobbing behaviour of the breeding chaffinches,* Fringilla coelebs. *Proceedings of the Royal Society of London B: Biological Sciences 269, 2345–2350. Pictures from Wikimedia Commons; chaffinch by tmwiki@gmail.com, owl by Martin Mecnarowski.*

2.4 STATIONARY ASSOCIATIONS CENTERED AROUND SPECIES INTERACTIONS

With some exceptions already mentioned, most of the aggregative phenomena described above are stationary. Here in this section, we also focus on generally stationary phenomena, but ones where the underlying cause for the group is the interaction between species, not an external factor.

2.4.1 Associations Based on a Protective Species

In Section 2.3.3, we discussed aggregations in predator-free space. In some cases, that predator-free space is maintained by a heterospecific partner. For example, animals place their nests near predators that are aggressive toward other predators (reviewed by Caro, 2005; Quinn and Ueta, 2008). In Section 5.1.4, we will talk about the complexities of these partnerships: usually commensal, they can occasionally become parasitic on the species providing the protection (Groom, 1992); in cases where benefits are bidirectional (Campobello et al., 2011), these kind of phenomena are essentially mixed-species colonies (Section 2.4.2).

Species that defend or guard their nests may also be used by other species. For example, some cichlids in Lake Malawi in Africa provide offspring defense for the first few weeks of life but then deposit their young near bagrid catfish, which pugnaciously defend their own broods and the cichlid young from predators that threaten them both (McKaye, 1985). The interaction is believed to also benefit the catfish because predators preferentially attack the cichlid fry, and the survival rate of the young catfish is thus higher when in association with cichlids. A similar phenomenon has been described in a marine fish that lays its eggs in the nest of another species that provides nest defense (Kent et al., 2011). In terrestrial environments, birds and a variety of invertebrates place their own nests near or inside the nests of ants, termites and wasps, which actively guard against disturbances (Roubik, 1988; Beier and Tungbani, 2006; Quinn and Ueta, 2008; Le Guen et al., 2015).

Interactions between species that maintain different levels of vigilance against predators are found repeatedly in MSAs generally, and in MSGs specifically. In an aquatic example, one species of stingray prefers to rest near another species that has a longer tail and hence potentially better predator detection (Semeniuk and Dill, 2006). Similar relationships may occur in mixed-species colonies (Section 2.4.2). For example, certain species of gulls and terns can detect predators quickly and pass along information to other species about them (Nuechterlein, 1981; Burger, 1984). In a related phenomenon, groups of mongooses associate with a variety of birds that provide vigilance and in return forage more efficiently on prey made available by the mongooses (e.g., Rasa, 1983; Flower et al., 2014). Because the groups move together (in and around the mongoose colony), we discuss them in the mixed-taxa MSG section of Chapter 3 (Section 3.6.3). The theme of vigilant and nonvigilant species associating in moving MSGs will be found repeatedly in other parts of Chapter 3, as well.

Protection also is a major benefit in many symbiotic relationships (Janzen, 1985), especially in those of sessile organisms, including marine animals, and invertebrates. In many cases these are strongly coevolved symbiotic systems in which the protected animal provides nutrition to its protector (e.g., lycaenid caterpillars and ants, Pierce et al., 2002; aphids and ants, Yao, 2014). In other cases, there are partial nutritional benefits (crabs that consume coral mucous and the algae that threaten to cover them, Stachowicz and Hay, 1999) or protection is bidirectional (sea anemones and anemonefish, Fautin, 1991). These kinds of associations usually involve interactions between two partner species at any one time or place, although many species may be involved throughout the world, with the overall association grading from obligate mutualism to parasitism (e.g., Pierce et al., 2002).

2.4.2 Mixed-Species Colonies or Roosts

We have encountered mixed-species nesting colonies already several times in this chapter in discussing enemy-free space and protective species. Such colonies are common in seabirds and wading birds. Roosts of birds or bats also can include multiple species coming together repeatedly in one place. A similar phenomenon may be found in marine mammals that rest together, including the stingrays mentioned above and dolphin species that rest together in shallow bays (Norris and Dohl, 1980a) although it is unclear whether the locations of these resting places are fixed. An invertebrate analogy is provided by whirligig beetles that form aggregations, some of which contain several species, when they are quiescent in the daytime, and which can be repeatedly found in the same location, day after day (Heinrich and Vogt, 1980).

These colonies have many similarities with MSGs; their movement patterns, pulsating away from and back to the colony, could be thought of as coordinated, and at a longer scale in time, you could think of the different species together establishing the colony and then dissolving it over weeks, months, or years. Yet the separation of foraging away from the colony or roost is distinctive and in general makes many foraging theories of MSGs (Chapter 4) not applicable to them. It is possible that information about food resources could be exchanged at the mixed colony or roost, as originally proposed by Ward and Zahavi (1973). In Section 4.2.1, however, we see that interspecific examples of information transfer of this kind are currently lacking.

The term "mixed-species colony" is also used for colonies of social web-building spiders. Participants in these colonies may share the costs of making web silk (Hénaut and Machkour-M'Rabet, 2010), use space that they would be unable to use by themselves (Hodge and Uetz, 1996), or capture prey that ricochet out of other species' webs (Uetz, 1989; Elgar, 1994; Section 4.2.2). These MSA could be mutualistic or commensal; in addition, there are other species of spiders that live in webs constructed by heterospecifics that are considered parasites (e.g., Grostal and Walter, 1997).

The spider example is part of a wider group of phenomena where species make constructions that provide shelter or habitat for other species. Because shelter can be seen as a form of protection, these are related to the kinds of protective species discussed above. These relationships can be primarily commensal, as in the many species that live in the burrows of prairie dogs (e.g., Agnew et al., 1986; although coinhabitants can provide vigilance for each other, Bryan and Wunder, 2014). Or, in the case of invertebrates that live in the nests of social insects (particularly ant, termite, and bee colonies), these associations can range the full gamut between parasites and mutualists (Hughes et al., 2008). In both these examples, animals may be clustered in space at one scale (in the constructions of one species) but not necessarily associated within the colony or nest. There are also close one-to-one symbioses in which shelter is traded for other benefits, such as that between burrowing shrimps that provide shelter to fish, which reciprocate with tactile signals of alarm (Karplus and Thompson, 2011).

2.4.3 Cleaning Mutualisms

Another association that is mostly stationary has two very different partners: animals that want to rid themselves of parasites and the generally smaller animals that eat those parasites. This is a species-rich mutualism especially in the ocean, where many fish species are serviced—even predatory fish, which when being serviced almost never prey on their cleaners—by a variety of fish and crustacean, which are sometimes obligatorily dependent on the activity (Côté, 2000). Usually the association is stationary, occurring at particular "cleaning stations" (Grutter et al., 2003). We discuss this marine phenomenon more in Section 5.4.4, especially the ways in which the "client" fish (those that want to be cleaned) can "punish" cleaners that "cheat" by eating client's mucous rather than the parasites.

In addition to the marine environment, some birds remove parasites, usually from large mammals but occasionally from lizards, turtles, or other birds (Sazima and Sazima, 2010). However, sometimes these interactions appear to be parasitic as birds may feed on wounds and actually prolong wound healing (Weeks, 2000; Section 5.4.3). As the mammals that these birds feed upon may move themselves, the overall interaction could have similarities to moving aggregations, discussed at several points in Chapter 3, and associations where one species makes prey more available for other species (Section 3.6). There are also some cleaning interactions among invertebrates. For example, some mites that live in bee hives (Section 2.5.1) provide cleaning services to bee larva (Biani et al., 2009).

2.5 CONCLUSIONS

In this chapter we have discussed a wide range of associations among animals. Some of these are groupings around resources or predators that we have termed aggregations (Section 2.3), and they are usually stationary. In others, the species interactions themselves are more important than external factors to form or maintain the group (Section 2.4), yet they are also stationary. In Chapter 3, we will focus on a wide range of moving phenomena: mostly MSGs but also some moving aggregations.

Among the phenomena discussed in this chapter, we also see a gradient in the asymmetry of benefits and in mutualism. Many aggregations result in competing taxa grouping together in close proximity, but we also described one aggregation type around predators—mobbing—in which participants mutually benefit. Stationary species interactions can range from commensalisms, like most of the examples of species nesting near protective species, to mutualisms, like those of cleaner fish and their hosts. The MSGs we discuss in Chapter 3 are primarily mutualisms, but some systems that consist of drivers and beaters are usually commensal, and some may be even parasitic on the leaders.

Our purpose in discussing this variety of MSAs is also to emphasize commonalities. Almost all associations can have an impact on predation pressure (Krause and Ruxton, 2002; Beauchamp, 2014). Theories that emphasize dilution of risk in groups or the transfer of information about predators will be applicable to many of the MSAs discussed in this chapter. Take, for example, one of the most iconic assemblages of species, ungulates coming together at a water hole on the African plains. In this situation the animals share information about predation (Périquet et al., 2010), because their predators are also quite aware of their aggregation. The complexities of interacting costs and benefits will be explored in greater detail in Chapters 4 and 5, often by exploring systems introduced in this chapter.

Two other patterns seen in this chapter will echo through other portions in this book. First, species have different traits that affect how useful they are to other species in MSAs. In this chapter, we discussed initiators of aggregations that are particularly efficient at finding resources or species that are particularly vigilant. Similar asymmetries in the benefits different species provide for MSGs are key to understanding species roles in them (Chapter 7). Second, although moving MSGs may be generally mutualistic, competition among species still can structure which species occur in such groups or where they are spatially located in the group. In this chapter, we discussed dominance hierarchies at stationary resources; similar hierarchies are important in structuring the MSAs that form around beating species (Willis and Oniki, 1978; Willson, 2004). Furthermore, even in moving bird MSGs, some species may be competitively excluded (Graves and Gotelli, 1993) or forced to suboptimal positions (e.g., Alatalo, 1981). Hence going forward, while maintaining a focus on facilitation, it is important to remember that competition can also structure the associations we discuss.

Chapter 3

Moving Mixed-Species Groups in Different Taxa

3.1 COMPARING MOVING MIXED-SPECIES GROUPS

Having reviewed the diversity of associations among animals of the same trophic level, we now turn to moving groups of species foraging together: mixed swarms, shoals, pods, herds, troops, and flocks. This chapter is organized by the frequencies these kinds of mixed-species groups (MSGs) are found with across taxa. We start with invertebrates, where the number of MSGs that have been described is small relative to the great diversity of species—with the caveat that the paucity of information for this taxon may be related to the definitions we choose or simply be due to a lack of study. We continue then with fish, which have usually been shown to prefer associating with conspecifics over heterospecifics, except in some habitats, such as coral reefs. Next, we consider mammals where MSGs are found frequently in certain social taxa, and finally birds where MSGs or moving aggregations are a dominant form of social organization in most habitats. The final section of this chapter focuses on a few moving groups where quite unrelated taxa interact together. We will review systems where one species acts as a beater or driver, making food more accessible for other species, several examples of cooperative hunting, and then a few associations between mongooses and other animals in which food accessibility is traded for antipredatory benefits.

Any scheme for classification of natural organisms ultimately fails due to the diversity of life, and so in some cases the distinctions between this chapter and Chapter 2 might be blurred. For example, in some cases described in this chapter—especially for birds foraging in water, in which the distribution of resources is difficult to detect—it is possible that the group is actually an aggregation over a resource, usually fish. Indeed, it is likely that some MSGs of large waders (Section 3.5.2) and seabirds (Section 3.5.4) are usually such kinds of aggregations. However, since the fish themselves move, we include them in this chapter to be able to compare them in their diversity and size to avian MSGs.

We organize each taxon's section by broad habitat types. Given the major role that predation plays in structuring MSGs (Chapter 5), classification by habitat is useful because different habitats can have very different predation regimes. This has been convincingly argued by Terborgh (1990) for birds and primates: animals that live in forest environments have short lines of sight and

Copyright © 2017 Elsevier Inc. All rights reserved.

must respond quickly to predators relying substantially on surprise; in contrast, grassland species have longer range visibility and can escape as a large, coordinated group. For a particular taxon, we will begin with aquatic habitats and particularly saltwater ones, and then work toward terrestrial habitats, ending with forested ones.

What information shall we compare among these different moving MSG systems? We first look at their composition: what families or genera tend to participate in them, or what ecological traits characterize participants. We refer to common names in the text, presenting scientific names in Table 3.1. We then describe the size of groups and their extent in the community: how many species participate in them? How prevalent are they in terms of the percentage of individuals of participating species that are found in them? How many species and individuals are typically found per group? A summary of representative values is provided in Table 3.1, a very "broad strokes" exercise because we ignore differences between studies in their definitions of a group, their duration of observations, and many other factors. We also use some admittedly awkward metrics, such as the midpoint of a range of species (see the header of Table 3.1), in an effort to wring out the most instructive data to compare across groups; mean numbers (such as the average number of species per group) were rarely available for some taxa, whereas ranges are so influenced by extreme values that they are not very informative.

In the text for each section on a particular taxon in a particular habitat, we concentrate on several further questions. We ask in what geographical regions groups form and in what seasons, or if there are any geographical or seasonal trends in group size. Furthermore, if there is information on the roles of species in flocks and community structure, we ask what taxa are involved in this or what ecological characteristics predict it. Finally, while our focus in this chapter is not on the adaptive benefits to grouping, we will point out certain theories that seem to apply especially to different taxa/habitats we discuss, or other characteristics that distinguish that kind of MSG from others.

3.2 INVERTEBRATES

As also noted in the last chapter, our definition of an MSG tends to bias the systems we cover against invertebrates, and our deemphasis of symbiotic associations (due to their persistence) is particularly important in this regard. Nevertheless, in researching for this book, it soon became apparent that descriptions of MSGs among invertebrates are scarce in the literature. It is unclear whether this pattern actually depicts the true distribution of MSGs in the invertebrates, perhaps associated with aspects of their biology, such as a reliance on species-specific pheromones. Alternatively, the small scale of the animals, and the difficulty of observations in the media that they are found in (e.g., soil, inside plants, in water), could be confounding variables. In particular, the scale complicates interpretation: for a very small animal, even a habitat patch is not

TABLE 3.1 The Number of Participating Species and Size of Mixed-Species Groups of Different Taxa and Habitats

Families or Genera Involved	References	Participating Species			Prevalence (%)			Midpoint of Range of Species			Midpoint of Range of Individuals		
		Range	Mean (SD)	N	Range	Mean (SD)	N	Range	Mean (SD)	N	Range	Mean (SD)	N
Invertebrate/Marine													
Crustacea: Mysida, Copepoda, Euphausiacea, Amphipoda; values for Mysid studies only	Wittmann (1977), McFarland and Kotchian (1982) and Ohtsuka et al. (1995)	2–16	8.3 (7.1)	3	Na	28.5 (Na)	1	2.0–4.5	3.3 (1.3)	2	125–2505	1315	2
Invertebrate/Forest													
Ants: Crematogaster and other genera	Orivel et al. (1997), Menzel and Blüthgen (2010) and Menzel et al. (2014)	2–6	3.6 (2.1)	3	43.5–74.0	58.8 (Na)	2	2.0–2.0	2.0 (0)	3	Na	Na	Na
Fishes/Marine													
Saltwater fish: diverse fishes, especially those abundant near coral reefs, such as Haemulidae, Mullidae, Labridae, Scaridae, Carangidae	Ehrlich and Ehrlich (1973), Alevizon (1976), Wolf (1985), Parrish (1989), Baird (1993), Sakai and Kohda (1995), Overholtzer and Motta (2000), Silvano (2001), Sazima (2002) and Sazima et al. (2007)	2–53	11.9 (15.6)	10	45.3–81.3	70.2 (23.4)	4	2.0–4.0	2.6 (0.8)	8	2–1750	262 (569)	9

Continued

TABLE 3.1 The Number of Participating Species and Size of Mixed-Species Groups of Different Taxa and Habitats—cont'd

Families or Genera Involved	References	Participating Species			Prevalence (%)			Midpoint of Range of Species			Midpoint of Range of Individuals		
		Range	Mean (SD)	N	Range	Mean (SD)	N	Range	Mean (SD)	N	Range	Mean (SD)	N
Fishes/Freshwater[a]													
Freshwater fish: diverse fishes, especially Cyprinidae and Gasterosteidae	Allan (1986), Allan and Pitcher (1986), Krause (1993a), Schlupp and Ryan (1996), Poulin (1999), Coolen et al. (2003), Mathis and Chivers (2003), Ward et al. (2003), Krause et al. (2005) and Camacho-Cervantes et al. (2013)	2–5	2.7 (1.0)	10	37.0–100	54.7 (30.9)	5	2.0–3.0	2.4 (0.4)	10	3–60	22.8 (23.7)	7
Mammals/Marine													
Dolphins: Delphinidae	Herzing and Johnson (1997), Scott and Cattanach (1998), Clua and Grosvalet (2001), Frantzis and Herzing (2002), Herzing et al. (2003), Acevedo-Gutiérrez et al. (2005), Maze-Foley and Mullin (2006), Quérouil et al. (2008), May-Collado (2010) and Kiszka et al. (2011)	2–12	3.6 (3.1)	10	1.4–100	28.6 (28.2)	10	2.0–3.0	2.2 (0.4)	10	11–1225	236 (417)	8

Families or Genera Involved	References	Participating Species			Prevalence (%)			Midpoint of Range of Species			Midpoint of Range of Individuals		
		Range	Mean (SD)	N	Range	Mean (SD)	N	Range	Mean (SD)	N	Range	Mean (SD)	N
Mammals/Grasslands													
Ungulates: especially Bovidae; also Equidae; rarely Giraffidae, Hippopotamidae	Gosling (1980), FitzGibbon (1990), Li et al. (2010), Kiffner et al. (2014), Pays et al. (2014) and Schmitt et al. (2014)	2–10	4.5 (3.1)	6	16.9–58.0	41.7 (19.1)	4	2.0–2.5	2.2 (0.3)	5	32–73	51.3 (16.9)	4
Mammals/Forests													
Primates: Callitrichidae; *Saimiri* and usually *Cebus*; Cercopithecidae	Gautier-Hion et al. (1983), Garber (1988), Podolsky (1990), Mitani (1991), Peres (1992), Bshary and Noë (1997), Fleury and Gautier-Hion (1997), Chapman and Chapman (2000), Regh (2006) and Pinheiro et al. (2011)	2–6	3.1 (1.4)	10	31.0–98.3	61.6 (22.7)	10	2.0–3.5	2.3 (0.6)	10	14–100	42.1 (30.8)	8

Continued

TABLE 3.1 The Number of Participating Species and Size of Mixed-Species Groups of Different Taxa and Habitats—cont'd

Families or Genera Involved	References	Participating Species			Prevalence (%)			Midpoint of Range of Species			Midpoint of Range of Individuals		
		Range	Mean (SD)	N	Range	Mean (SD)	N	Range	Mean (SD)	N	Range	Mean (SD)	N
Birds/Marine[b]													
Seabirds: Primarily Laridae, Alcidae, Procellariidae, Diomedeidae, Sulidae	Porter and Sealy (1981), Grover and Olla (1983), Harrison et al. (1991), Ballance et al. (1997), Camphuysen and Webb (1999), Ostrand (1999), Maniscalco et al. (2001), Silverman and Veit (2001), Zamon (2003) and Anguita and Simeone (2015)	5–49	18.7 (12.8)	10	6.8–57.2	20.4 (24.6)	4	3.5–5.0	4.6 (0.8)	4	26–545.5	167.7 (198)	6
Birds/Coastal													
Shorebirds: especially Scolopacidae and Charadriidae; sometimes Laridae	Ashmole (1970), Stinson (1980), Byrkjedal and Kålås (1983), Thompson and Barnard (1983), Metcalfe (1984), Stawarczyk (1984), Byrkjedal (1987), Battley et al. (2003), Cestari (2009) and Gavrilov (2015)	2–18	7.2 (5.3)	10	12.9–53.2	29.0 (19.0)	4	2.0–6.5	2.8 (1.8)	6	2–135.5	68.9 (62.2)	5

Families or Genera Involved	References	Participating Species			Prevalence (%)			Midpoint of Range of Species			Midpoint of Range of Individuals		
		Range	Mean (SD)	N	Range	Mean (SD)	N	Range	Mean (SD)	N	Range	Mean (SD)	N
Birds/Freshwater													
Waterfowl: mostly Anatidae, some Rallidae, Podicipedidae	Anderson (1974), Bailey and Batt (1974), Byrkjedal et al. (1997), Källander (2005), Kristiansen et al. (2000), Larsen (1996), Paulson (1969), Pöysä (1986b), Randler (2004) and Silverman et al. (2001)	2–12	4.8 (3.6)	10	2.0–90.1	34.9 (30.5)	6	2.0–3.5	2.3 (0.6)	8	2–96	21.8 (32.3)	8
Birds/Wetlands													
Waders: Ardeidae, Threskiornithidae, sometimes Phoenicopteridae	Emlen and Ambrose (1970), Kushlan (1977), Willard (1977), Russell (1978), Caldwell (1981), Erwin (1983), Master (1992), Smith (1995), Bennett and Smithson (2001) and Kyle (2006)	3–20	9.4 (6.3)	8	65.7–83.2	74.0 (Na)	2	3.0–6.1	4.6 (1.5)	5	76.5–302.5	150.6 (75.6)	7

Continued

TABLE 3.1 The Number of Participating Species and Size of Mixed-Species Groups of Different Taxa and Habitats—cont'd

Families or Genera Involved	References	Participating Species			Prevalence (%)			Midpoint of Range of Species			Midpoint of Range of Individuals		
		Range	Mean (SD)	N	Range	Mean (SD)	N	Range	Mean (SD)	N	Range	Mean (SD)	N
Birds/Grasslands													
Grassland birds: diverse birds, especially Emberizidae, Fringillidae	Cody (1971), Rubenstein et al. (1977), Greig-Smith (1981), Barnard and Stephens (1983), Clergeau (1990), Zamora et al. (1992), Veena and Lokesha (1993), Rolando et al. (1997), Ridley and Raihani (2007) and Canales-Delgadillo et al. (2008)	2–29	10.8 (9.8)	10	11.0–71.1	27.7 (23.1)	6	2.0–7.0	3.1 (1.7)	8	4.5–125	32.8 (42.4)	7

Families or Genera Involved	References	Participating Species			Prevalence (%)			Midpoint of Range of Species			Midpoint of Range of Individuals		
		Range	Mean (SD)	N	Range	Mean (SD)	N	Range	Mean (SD)	N	Range	Mean (SD)	N
Birds/Forests													
Forest birds: diverse range of Passeriformes; also Picidae, Trogonidae, Certhiidae, some Cuculidae	Vuilleumier (1967), Austin and Smith (1972), Partridge and Ashcroft (1976), Gradwohl and Greenberg (1980), Morrison et al. (1987), Dean (1990), Eguchi et al. (1993), Mönkkönen et al. (1996), Chen and Hsieh (2002) and Arbeláez-Cortés et al. (2011)	4–75	23.9 (20.8)	10	13.4–72.1	43.4 (22.2)	6	3.0–12.5	6.1 (3.0)	8	10.5–65	31.2 (20.4)	6

Note that this course resolution exercise does not incorporate differences between studies in their definitions of a group or in the duration of their observations. "Participating species" are the number of species that were recorded participating in all observed MSGs. "Prevalence" represents the percentage of observations of participating species in which they were in MSGs, averaging across species, or in some cases, across sites. The "midpoint of range" of species or individuals is the mean between the least number reported and the most recorded; the range of such values are shown for the sample of papers. Using a range was chosen over using a mean because ranges were much more often recorded; if means were given, we estimated the range as the mean ± 2SD. If there were more than 10 papers per taxa, they were selected randomly. Na = Data not available.

[a] Of the 10 studies, 7 were conducted in an experimental situation, and these values might be quite unrepresentative of the field situation.

[b] For the seabirds, we avoided including long-term (in the order of more than several hours) "Type II" or "Type III" aggregations.

very large. So, for instance, the aggregation of several pests and pathogens on a plant at one time, with interacting effects (e.g., Wertheim et al., 2005), is analogous to vertebrates clustering in a habitat (and therefore not an MSG), although the individual insects may be very close to each other when measured in meters. Another example of extremely complex species interactions in a small space is that presented by microorganisms in animals' digestive systems (Brune and Friedrich, 2000; Eckburg et al., 2005).

Regardless, we have encountered descriptions of at least two kinds of MSGs of invertebrates that meet our requirements here, especially that of movement. The first are marine groups of crustaceans. The best described such "swarms" are mysids in temperate or subtropical shallow seas (Fig 3.1; Wittmann, 1977; Ohtsuka et al., 1995). Less well described are MSGs of copepods (Ueda et al., 1983), euphausiids, and amphipods (Fenwick, 1978). Mysids may even associate with post-larval fish of the same size (McFarland and Kotchian, 1982). In all crustacean MSGs, the overall frequency of MSGs compared to monospecific ones is low. For example, in mysids, most groups are composed of one species and when MSGs do form, the majority of the individuals are of one species, and there are only a few individuals of other "guest" species (Wittmann, 1977; Ohtsuka et al., 1995). Little is known about the roles of species in these groups or geographic or seasonal variation. In terms of adaptive benefits to grouping, as Wittmann (1977) first suggested for mysids, when there are only a few individuals at a certain place and time, they may choose to join another species' group, rather than be isolated outside a group.

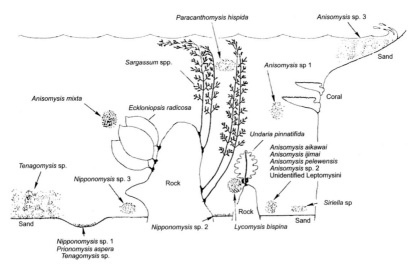

FIGURE 3.1 Mysid groups in shallow seas off Japan and Korea. A minority of groups are formed by more than one species; however, usually one species dominates numerically. *Used by permission from Ohtsuka, S., Inagaki, H., Onbe, T., Gushima, K., Yoon, Y.H., 1995. Direct observations of groups of mysids in shallow coastal waters of western Japan and southern Korea. Marine Ecology Progress Series 123, 33–44.*

Swarms are also found in flying insects. In mosquitoes, such as mysids, a small minority of swarms may include more than one species (Sawadogo et al., 2014). However, such swarms are formed for mating purposes, and hence are quite different from the other phenomena described in this chapter (Lindeberg, 1964).

In the ants, there are a few "parabiotic" species in which ants of two different species both live in the same nests and forage together (Orivel et al., 1997; Menzel and Blüthgen, 2010), using the same pheromone trails (Menzel et al., 2010). These associations often include ants of the genus *Crematogaster* interacting with ants of several other genera. Experiments seem to indicate that *Crematogaster* ants are better at finding resources, whereas ants of the other genera provide nest defense in return (Menzel and Blüthgen, 2010), but in other studies the benefits to *Crematogaster* were not obvious (Menzel et al., 2014). The close association of these ants in the foraging context makes them similar to other phenomena of this chapter; however, this is a relationship between just two species in any one place and is generally classified as a symbiosis.

3.3 FISH AND AQUATIC AMPHIBIANS

3.3.1 Saltwater Fish

The ideas discussed by Terborgh (1990) above about the importance of habitat in determining MSG organization may also apply to fish (Ashley Ward, personal communication). In the pelagic environment with high visibility and no substrate for protection, highly coordinated fish "schools" form, and in general, these tend to be monospecific, or occasionally formed of highly similar species. However, in environmentally complex nearshore areas, fish form less coordinated groups termed "shoals," and MSGs may be fairly common. Nevertheless, similar to the mysid groups discussed earlier, fish species in such nearshore MSGs may not be mixed together fully, because they tend to segregate by species, with different species inhabiting different depths (Parrish, 1989) or forming monospecific clusters within the larger group (Ehrlich and Ehrlich, 1973). Some marine MSGs may also dissolve in the presence of predators (Wolf, 1985). In terms of their distribution, marine fish MSGs are particularly common for juvenile fish that may be quite vulnerable to predation (Overholtzer and Motta, 2000). The systems that have been studied so far are tropical (13 of 13 empirical studies we know of), but this may represent a research bias, because MSGs can also commonly be found in temperate waters near Sydney, Australia (Ashley Ward, personal communication).

A particular kind of marine fish MSG is found in coral reef environments, between "nuclear" species that disturb the substrate and those that follow them (Lukoschek and McCormick, 2000). The nuclear species can be any species that makes such disturbance, including nonfish such as octopi and turtles, but the followers are almost always fish (Sazima et al., 2007). Indeed, although at any one

(A)   **(B)**

FIGURE 3.2   Two kinds of mixed-species fish groups in a coral reef off Brazil. In (A), the morphologically dissimilar *Anisotremus virginicus* (*white circle*) join a group of *Haemulon auro-lineatum*, joining this abundant species perhaps because they do not have enough numbers to form their own group. In (B), *Mulloidichthys martinicus* (*white circle*) joins *H. aurolineatum*. This species looks very similar to the group they join, an example of apparent "protective mimicry," whereby the color pattern helps them successfully join a group and not be easily targeted as different from the rest of the group by a predator. *Adapted from Pereira, P.H.C., Feitosa, J.L.L., Ferreira, B.P., 2011. Mixed-species schooling behavior and protective mimicry involving coral reef fish from the genus* Haemulon *(Haemulidae). Neotropical Ichthyology 9, 741–746 which was published under a Creative Commons license.*

time only a few species, and often just two, participate, the association might be quite prevalent in the community, with as many as 20% of fish in a coral reef environment participating at some point (Table 3.1; Sazima et al., 2007). In general, the following species in these relationships is the one that benefits. In some cases, however, the relationship between species may not be so simple, with both species attracted to each other and more mutualistic benefits (Baird, 1993). As with other tropical systems, seasonal variations are expected to be minor; however, juvenile fish could behave differently than adults at different stages of their development.

MSGs of coral reef fishes can also be related to mimicry (Section 6.3.3). Aggressive mimicry is a kind of mimicry found in coral reef fish, wherein one carnivorous species imitates a harmless one, thereby gaining access to food and particularly the ability to catch other unwary fish (Sazima, 2002). As many as 60 species worldwide have been reported to use this tactic (Moland et al., 2005). Many of the species that show such imitation group with the modeled species, as this makes the mimicry more effective. Another kind of "social" or "protective" mimicry is when two species with very similar color patterns group together, presumably because they enjoy mutual protection from predators (Moland et al., 2005; see an example of potential protective mimicry in the bottom of Fig. 3.2, compared to the same species associating with morphologically distinct fishes on the top of that figure). This type of mimicry has been suggested particularly for scale-finned grunts, a genus that is particularly abundant on coral reefs (Krajewski et al., 2004; Pereira et al., 2011).

3.3.2 Freshwater Fish and Aquatic Amphibians

Freshwater fish MSGs can be relatively common in the wild when the species are abundant (e.g., Allan and Pitcher, 1986; Coolen et al., 2003), especially among certain common families and in shallow waters and littoral zones (Table 3.1). In general, however, freshwater fish seem to prefer to join conspecific groups more than MSGs, when given the choice in the laboratory. Only in certain experimental conditions will fish prefer heterospecifics over conspecifics. For example, fish prefer familiar heterospecifics over unfamiliar conspecifics (Ward et al., 2003), or heterospecifics that were fed the same diet over conspecifics fed a different diet (Kleinhappel et al., 2016). Another experiment has shown that fish will join another species that is preferred by predators under high predation pressure (although still preferring conspecific shoals when predation pressure is low; Mathis and Chivers, 2003; Section 5.1.5). Mollies will choose a group of three heterospecifics over a lone conspecific but will select conspecifics at lower heterospecific numbers (Schlupp and Ryan, 1996). Similarly, a guppy joins heterospecifics when its own numbers are low and the choice is to school with heterospecifics or be alone (Camacho-Cervantes et al., 2013). Generally, little is known about leadership or geographical and seasonal variation in these groups because most of the work in these species has been conducted in the laboratory rather than in the wild.

The preference of freshwater fish for conspecifics may be related to fish choosing to join groups in which the other fish are similar in length and preferred swimming speed, a pattern that can be seen both within and between species (Krause and Godin, 1994; Ward and Krause, 2001; Ward et al., 2002). This preference may in turn be due to the "oddity effect," in which fish that are in some way different from others in a group are preferentially selected by predators (Landeau and Terborgh, 1986; Section 5.1.3). Indeed, when faced with a predator, some freshwater fish MSGs break apart (Allan and Pitcher, 1986), similar to what happens in saltwater MSGs. A preference for similarly sized fish could also be due to smaller fish avoiding competition from larger ones (Ward et al., 2002).

An interesting phenomenon related to grouping in freshwater communities is air breathing, where animals in anoxic conditions come to the surface to breathe air synchronously, probably as an adaptation to reduce predation risk at the surface (Kramer and Graham, 1976). In some anoxic pools, fish of different species could also synchronize their air breathing together (personal communication, the late Jeffrey Graham). Aquatic adult frogs may also participate in synchronous air breathing (Baird, 1983), although it is not known if there is a heterospecific component to such behavior.

The earlier discussion of frogs emphasizes how predation risk may be similar for aquatic phases of amphibians as for fish. As mentioned in the last chapter, tadpoles in Madagascar group together in clear ponds (Glos et al., 2007a), and further work needs to study tadpole communities in other regions.

3.4 MAMMALS

The overall impression of mammalian MSGs, in comparison to birds, discussed below, is that, while there are examples of very strong interspecific relationships, including some species pairs that live in association much of their lives (Garber, 1988), the proportion of mammalian species that participate in MSGs is low. The main reason behind this is that the most diverse mammalian taxa (bats and rodents) are not highly social (with the exception of bat-roosting behavior, a topic which fits best in the last chapter about aggregations in predator-free areas; Rainho and Palmeirim, 2013). MSGs in mammals occur mostly in species that live in conspecific groups (Stensland et al., 2003). Hence the primary taxa include marine mammals, ungulates, and primates. As none of these taxa have high numbers of species in any one place, mammalian MSGs often have a relatively low number of species involved (Table 3.1).

3.4.1 Cetaceans

Mammalian MSGs of the oceans are largely comprised of dolphins. Like the aquatic invertebrates and fish we have encountered so far in this chapter, dolphin MSGs are a minority of the total number of dolphin groups (for 9 of 10 studies included in Table 3.1, less than 36% of observations included more than one species). In some areas, two or more species of dolphins are common and sympatric but never seen in MSGs (Bearzi, 2005). Similar to other mammalian MSGs, species richness is low, with groups of three or more species being exceedingly rare (see Table 3.1; for examples of systems with more than two species, see Clua and Grosvalet, 2001; Frantzis and Herzing, 2002).

Generally, the size of dolphin MSGs is larger than monospecific groups, as the pods of two species fuse together, and in individual numbers they appear to be among the largest MSG systems known (Table 3.1). The actual size of MSGs varies by where they are encountered. Groups close to sea shore may average less than 15 individuals (Herzing and Johnson, 1997), whereas groups in the open ocean can include hundreds to thousands of animals (Scott and Cattanach, 1998; Gowans et al., 2008). The reasons behind such differences are similar to the drivers of group size in monospecific dolphin groups: unpredictable resources, large ranges, and high predation pressure predict the large size of oceanic groups (Gygax, 2002; Gowans et al., 2008; Cords and Würsig, 2014). These large oceanic dolphin MSGs are often associated with tuna and seabirds (Au and Pitman, 1986; Scott and Cattanach, 1998; Section 3.6.1), and sometimes similarly large groups of animals can occur in upwelling zones close to the coast, where they fit more closely the idea of aggregations than moving pods (Clua and Grosvalet, 2001). Another important characteristics of dolphin MSGs is that as "fission-fusion societies," individuals can change group membership rapidly over time, and dolphin MSGs have rather rapid changes in composition, quite different from primates (Cords and Würsig, 2014).

The question of leadership and following in these associations, and their initiation, is difficult to assess for humans above the surface (Norris and Dohl, 1980b; Scott et al., 2012). However, in three studies, there is a trend for rarer species to be seen more exclusively in MSGs (Frantzis and Herzing, 2002; Psarakos et al., 2003; Quérouil et al., 2008). This suggests that these rare species follow other species to increase their group size. Another feature of dolphin MSGs is a strong element of interspecific aggression, with larger species chasing, attacking, or sexually harassing smaller species (Herzing and Johnson, 1997; Psarakos et al., 2003; May-Collado, 2010). Indeed, hybridization is sometimes found in these species and is thought to be a consequence of this sexual aggression (Herzing and Elliser, 2013). For these reasons, smaller subordinate species may gain less from MSGs than dominant ones (Clua and Grosvalet, 2001).

As they are found primarily, though not exclusively, in tropical seas (Au, 1991), it is probable that dolphin MSGs occur year-round, although this requires further study. Also unclear is how persistently dolphins interact with other species such as dugong (Kiszka, 2007) or baleen whales (Shelden et al., 1995), and what the adaptive benefits of such associations are, if any.

3.4.2 Ungulates

The most commonly reported mammalian MSGs in grassland environments are those composed of ungulates (FitzGibbon, 1990; Kiffner et al., 2014). Ungulate MSGs may include two closely related species, such as two species of gazelle (FitzGibbon, 1990; Li et al., 2010), or species with widely different morphology, such as zebras and wildebeest (Sinclair, 1985; Kiffner et al., 2014; Fig. 3.3), as long as the species still share common predators (Schmitt et al., 2014). Like cetaceans, these tend to be low-diversity systems, generally with just two species interacting. Other grassland mammals that can be found in MSGs

FIGURE 3.3   Zebra and wildebeest grazing together in Tanzania. The species are often found together (Kiffner et al., 2014), with zebras usually on the leading edge of a wildebeest group (Sinclair, 1985). *Picture from Wikimedia Commons, created by Muhammad Madhi Karim.*

include kangaroos, though such interactions appear rare, and only between closely related species (Coulson, 1999).

These herbivorous species can have different preferences for the height of grass they consume (Sinclair, 1985) and may interfere with each other when foraging together (Pays et al., 2014). Due to these factors, ungulate MSGs may have stronger associations when food resources are not limiting, such as in the wet season (Gosling, 1980; Kiffner et al., 2014). These MSGs may also be seasonal because of migration patterns (Sinclair, 1985) and may vary in their sexual composition throughout the year, as do monospecific ungulate herds (Ruckstuhl and Neuhaus, 2002; Li et al., 2010). When sexes are mixed, a cost to ungulate MSGs may be hybridization (Li et al., 2010), similar to dolphins.

Nuclear species for ungulate MSGs have been described as species that are especially vigilant (Sinclair, 1985; Pays et al., 2014; Schmitt et al., 2014). As grazing animals, ungulates by and large eat widespread plant species and hence the benefits of foraging together, such as efficiency in finding food, would seem to be limited (Sinclair, 1985); therefore, antipredatory benefits to MSGs would seem dominant. Indeed, recent studies have documented a "mixed-species herd effect" in which individual animals devote less time to vigilance in the presence of heterospecifics, controlling for the size of the group (Scheel, 1993; Kluever et al., 2009; Périquet et al., 2010; Pays et al., 2014; Schmitt et al., 2014). It has also been reported that some species might choose to join species that predators prefer, to deflect predation risk on to others (FitzGibbon, 1990; Schmitt et al., 2014).

3.4.3 Primates

Primate MSGs have been extensively studied, with at least 60 articles to our knowledge, making this taxa second only to forest birds in a number of MSG publications. In contrast to birds, however, primate MSGs are very limited in their distribution, both taxonomically and geographically. We see three well-documented systems: (1) groups of tamarins in South America (Heymann and Buchanan-Smith, 2000), (2) groups that include spider monkeys in South and Central America, which associate often with capuchins (Pinheiro et al., 2011), and (3) associations in Africa between a wide range of arboreal forest monkeys (Struhsaker, 1981) and even the terrestrial mandrill (Astaras et al., 2011). In general, most primate troops contain just two species; in a minority of observations, this can be the result of just one individual incorporating itself in a heterospecific group (Fleury and Gautier-Hion, 1997). However, usually the groups of the two species fuse, like dolphins, resulting in a fairly high number of individuals per group (Terborgh, 1990; see Table 3.1). A few systems regularly have three species (e.g., Regh, 2006), or occasionally more (Mitani, 1991).

The stability of primate MSG systems ranges strikingly: this gradient runs from two species of tamarins forming groups that are almost always together and mutually defend a common territory (Garber, 1988) to haphazard encounters between monkeys at fruiting resources (Waser and Case, 1981). Generally, because these MSGs are exclusively tropical, there is not a great deal of seasonal variation, although there can be differences between wet and dry seasons, perhaps related to the distribution of resources (Regh, 2006).

Stability also varies geographically: for example, tamarins are not found in MSGs north of the Amazon river (Heymann and Buchanan-Smith, 2000), and spider monkeys, which can be found more inside of MSGs than out of them in South America (Terborgh, 1983), may only participate less than 10% of their time in MSGs in Central America (Boinski, 1989). In Africa, associations between monkeys are strong in West Africa, with some species having completely overlapping territories and "near-permanent" associations (Eckardt and Zuberbühler, 2004). But most associations in parts of eastern Africa, such as Kibale forest in Uganda, are weaker and do not pass statistical tests of random association (e.g., Waser, 1984). Even within an area of 800 km^2, associations can vary greatly in their frequency, which could be related to varying predation risk (Chapman and Chapman, 2000). The same argument is made to explain the lack of primate MSGs in Asia (but see Bernstein, 1967) and Madagascar (but see Freed, 2007), where monkey-eating eagles are rare or absent.

The subject of leadership has not been as much discussed in primate MSGs as it had been in birds. In stable associations between mustached and saddleback tamarins, mustached tamarins consistently lead (Smith et al., 2003), perhaps related to their dominance over the other species. In Africa, Diana monkeys are thought to be central to MSGs because of their loud and far-traveling calls associated with predation threats (Bshary and Noë, 1997; Buzzard, 2010). Another study showed that red-tailed monkeys actively initiated, maintained, and terminated associations with red colobus, apparently to take advantage of the colobus' vigilance (Teelen, 2007).

Niche separation among species has emerged as one general feature of primate MSGs. Most participants in these systems are omnivorous or frugivorous, and thus, because of the clumped nature of fruiting trees, competition might be stronger than in forest bird MSGs, which are mostly insectivorous. Indeed, aggressive interactions between species in primate MSGs tend to occur in fruiting trees (Terborgh, 1990). More intense competition could drive niche differentiation. For example, in tamarins, species join MSGs only if they are more than 8% different in their size (Heymann, 1997), and species in groups are very clearly different in their vertical stratification (Heymann and Buchanan-Smith, 2000). African primate MSGs also are strongly vertically stratified (Gautier-Hion et al., 1983). Such differences in the microhabitats among MSGs members can have several consequences: reduced competition, niche expansion, and also increased collective vigilance of the MSG, as both aerial and terrestrial threats can be detected more easily (Peres, 1993; McGraw and Bshary, 2002; Stojan-Dolar and Heymann, 2010; Section 5.1.2). While associations between primates that have very similar food preferences are rare, when they do occur the associating species have complementary antipredator capacities (Eckardt and Zuberbühler, 2004).

3.5 BIRDS

We now turn to birds, which are clearly the taxa most often found in MSGs specifically and in mixed-species associations (MSAs) more generally. With the exception of birds of prey and large flightless birds such as ostriches, almost all other bird taxa are found at some point in MSAs, at least in the nonbreeding season, when groups are usually more frequent, following the dissolution of nesting territories. As bird diversity at any one area is high, there are regularly more than two species per group. For example, in Table 3.1, the only taxa with "midpoint ranges" of species per group greater than 3.5 are birds, with the highest extreme occupied by forest birds (6.0). Following the organization of the chapter so far, we start with waterbirds, which are less stable, and save the last section for the most stable, diverse MSGs in forests.

As discussed briefly above, waterbirds provide a problem of discriminating between aggregations built around resources, which, in the case of fish, may move, and moving groups. It is easy to think of gulls obtaining discards from a trawling boat as an aggregation (Hudson and Furness, 1988), but are clusters of sandpipers, herons, ducks, or seabirds aggregations or groups? As noted in the last chapter, the only real way to distinguish MSAs from MSGs is to look at correlations between the resources underwater and the birds, which is rarely done (but see Veit et al., 1993; Burger et al., 2004). Different groups of species living in different ecosystems and experiencing varying rates of predation have more or less attributes of "moving flocks," and we will thus examine them separately (Sections 3.5.1 through 3.5.4); for waders, waterfowl, and seabirds we will refer to them as MSAs because their feeding resources are clumped. Note that two of these taxa (wading birds and seabirds) also nest colonially. For discussion of their colonial nests, please see Section 2.4.2; here, we concentrate on interactions between the species when feeding.

3.5.1 Seabirds

In this section we will concentrate on short-term MSAs among marine birds that occur when fish come close to the ocean surface. Such seabird MSAs, termed "Type I" aggregations by Hoffman et al. (1981), can be finished in a matter of minutes. Longer-lasting associations, referred to as "Type II" or "Type III" aggregations, also occur, whether near trawling boats as discussed above, or at fronts (boundaries between water masses; Schneider, 1982), upwellings, or other areas where fish are massed for several hours or even days (Duffy, 1983, 1989; Hunt et al., 1988). Indeed, as predation seems to be low for seabirds, almost all marine bird MSAs have been linked to resource distribution. The shorter-lasting interactions, however, have some similarities with moving MSGs, in that the attraction to the association may be primarily driven by new recruits cueing on the behavior of birds already there.

The roles of species in these MSAs are critical to their functioning. That seabird MSAs are assembled by birds responding to certain initiating or "nuclear" species was first remarked on by Sealey (1973) but elaborated in a classic article by Hoffman et al. (1981), who called such species "catalysts," and also classified other species as "divers," "kleptoparasites," and "suppressors." Catalysts are usually gulls in northern temperate systems, which by circling after their first hunt of a fish, indicate that there are more fish present (Hoffman et al., 1981; Chilton and Sealy, 1987). Divers, such as alcids and cormorants, cue off the catalysts and may quickly approach a developing MSA. Because of high visibility in the open ocean environment, birds might be able to cue off each other up to 4 km and beyond and arrive from those distances within 5–10 min (Haney et al., 1992). Kleptoparasites are jaegers or gulls that attack other species, especially those that have stored fish in their gullets for their nestlings; high activity of such species in the center of the flock may push other species to the margins. Finally, suppressors are species whose activities lead to the end of the interaction by decreasing the number or the availability of the fish. For example, shearwaters may plunge-dive in large numbers into the fish school, either decimating their numbers or sending them deeper into the water column (Hoffman et al., 1981), and kleptoparasitic species such as herring gulls can also bring MSAs to an end (Camphuysen and Webb, 1999).

Diving bird species can also play a more central role in chasing fish toward the surface or forcing them into tight balls that are then accessible to other species. Grover and Olla (1983) first described the role of alcids in driving fish, and later articles such as Chilton and Sealy (1987), Mahon et al. (1992), and Ostrand (1999) showed that many "Type I" MSAs were actually initiated by divers driving fish toward the surface. Hence, these species might better be termed "initiators" or "producers" of MSAs (Camphuysen and Webb, 1999).

Beyond the complementarity of species roles in seabird MSAs, competition may also shape the composition of groups. Larger species may be able to displace smaller ones from the preferred places in groups, as has been reported in gulls (Camphuysen and Webb, 1999) and terns (Safina, 1990), and such competition may shape large-scale spatial variation in MSAs, by structuring which species are able to monopolize groups in high-resource areas (Ballance et al., 1997). Recently, social network analyses have indicated that associations form between species of the same body size, perhaps because of smaller species avoiding dominants (Anguita and Simeone, 2016).

The description above about seabird MSAs is one based on northern temperate systems. In the Antarctic, penguins and/or fur seals drive fish toward the surface. Black-browed albatrosses have been observed to be catalysts, acting as cues to other seabirds (Harrison et al., 1991; Silverman et al., 2004). In the tropics, such MSAs are particularly dependent on predatory fish and dolphins to bring prey fish to the surface (Au and Pitman, 1986; Ballance et al., 1997; Section 3.6.1). Tropical seabird MSAs may continue year-round despite major changes in the availability of food (Anguita and Simeone, 2015). Temperate systems are likely to be seasonal because the abundance of seabirds may be lower in winter (Burger et al., 2004), but year-round studies are scarce. Another important gradient of geographical variation is from the coast to far offshore, with the composition and frequency of MSAs changing quite dramatically (Camphuysen and Webb, 1999). Some systems may be concentrated very close to the coast (as in the flocks described by Mills, 1998; less than 60 m offshore of the Galapagos), whereas others can be thousands of kilometers off the coast (Ballance et al., 1997).

3.5.2 Shorebirds

Perhaps the easiest aquatic birds to count as MSGs rather than MSAs are the shorebirds, which usually inhabit coastlines but sometimes can be found in agricultural fields (Fig. 3.4). These MSGs primarily include plovers and sandpipers, although they can also be joined by gulls, which may steal resources from the other birds (Barnard and Thompson, 1982). Sandpipers and plovers do not feed on clumped food (and hence these systems are almost certainly not aggregations), but rather their movement is most influenced by the tides (Burger et al., 1977b) and they often fly or rest together (Stinson, 1980). Sometimes, as predicted by Terborgh (1990) for groups in open habitats, they form between species that look very similar, such as between species of small *Calidris* sandpipers (Cestari, 2009), and some research has suggested that similarities in plumage may reduce interspecific aggression (Stawarczyk, 1984). Nevertheless, many sandpipers that do not join MSGs or that are found in other geographical areas also look similar, suggesting inheritance of a similar plumage from a common ancestor (Beauchamp and Goodale, 2011). Shorebird MSGs may also be composed of different kinds of shorebirds that look very different from each other, such as sandpipers and plovers (Byrkjedal and Kålås, 1983; Gavrilov, 2015).

FIGURE 3.4   A mixed-species group of shorebirds in California. The group consists of willets and marble godwits. *Picture from Wikimedia Commons, created by Ingrid Taylar.*

Shorebird MSGs are generally of two species. Yet, despite the prediction for birds in such open habitats to be in large group sizes (Terborgh, 1990), in practice the number of individuals is often quite small (e.g., between 3 and 10 birds in the observations of Gavrilov, 2015), perhaps because of the costs of foraging together such as interference. Shorebird MSGs can be found during migration (Gavrilov, 2015), on wintering grounds (Cestari, 2009), and breeding grounds (Byrkjedal and Kålås, 1983).

The overall threat to this group of predation by raptors in high (Page and Whitacre, 1975), and these MSGs seem organized around vigilance. Birds will quickly respond to the alarm flight of other species (Metcalfe, 1984). Birds fleeing from a predator form swerving, or tightly packed and dense flocks (Michaelsen and Byrkjedal, 2002). The oddity effect may be relevant to such flocks, making similar species more liable to flock (Terborgh, 1990). The organization and assembly of the MSGs is asymmetrical with some species that are not as visually vigilant associating with more vigilant species and then scanning less and feeding more (Byrkjedal and Kålås, 1983; Thompson and Thompson, 1985; Gavrilov, 2015). Despite the costs associated with their kleptoparasitic activity, gulls may be particularly vigilant (Thompson and Barnard, 1983).

3.5.3 Waterfowl

Waterfowl MSAs include ducks and geese, as well as some members of the rail and grebe families. While sometimes these are found off coasts, the majority of observations have been in freshwater. Waterfowl groups tend to be loose and of few species, with most observations of two species only (Pöysä, 1986b). They can be found throughout the year: in the breeding season (Pöysä, 1986b), in wintering grounds (Schummer et al., 2008), and during migration (Silverman et al., 2001).

A large part of the reports on waterfowl MSAs is related to some species providing a disturbance that facilitates food capture for the others. For example, with their long necks, swans are able to feed on bottom vegetation, and their disturbance brings vegetation to the surface that dabbling species can use (Section 4.2.2). Groups that form around swans are perhaps the most species-rich waterfowl MSAs, with reports of an average of more than three species in addition to the swans themselves (Källander, 2005). Other systems in which some species beat up resources that others use include scoters followed by grebes (the scoters bring benthic invertebrates to the surface that grebes consume, Byrkjedal et al., 1997), or large shorebirds following ducks (Jacobsen and Ugelvik, 1994).

Waterfowl predation pressure is particularly high when they are outside of water, and ducks or geese grazing in fields have been often found to lower their individual levels of vigilance in MSAs, at least for some of the participating species (Larsen, 1996; Kristiansen et al., 2000; Randler, 2004; Jonsson and Afton, 2009). In the water, foraging may be more of a focus, and in addition to the following interactions discussed above, some species may converge on each other's niche, benefiting from copying the successful foraging of other species, whereas competitive interactions may drive other species' niches apart (Pöysä, 1986a).

3.5.4 Waders

These MSAs consist of larger wading birds than the shorebirds that prefer to forage in marshes near coastlines or in wet savannas and are generally composed of taxa that are fish-eating, such as the herons and the ibis and spoonbills. Except in certain situations (Caldwell, 1986), predation is low on these large birds. Rather, they are generally explained as aggregations formed where fish are plentiful, such as in pools in estuaries where the water has been receding (Kushlan, 1977; Fig 3.5). There is an element of heterospecific attraction, however; experiments with model herons have shown that just the presence of particular species attracts more birds (Kushlan, 1977; Caldwell, 1981; Master, 1992; Green and Leberg, 2005; Section 6.2.1). Unlike seabirds, however, where aggregations can last only minutes, and thus there is an ever-continuous movement between groups, these wader MSAs tend to last longer, in the range of several hours, and perhaps most of the day (Master, 1992). These long-lasting aggregations can also include gulls and terns (Master, 1992; Schreffler et al., 2010). Because of the distribution of the bird species, they are generally tropical; however, they can occur during the breeding season in more temperate regions.

Like other aggregative systems, which we mostly discuss in Chapter 2, wader MSAs tend to form around cores of monospecifically gregarious species. Furthermore, dominant species may drive subordinate ones from preferred places in the group (Russell, 1978). In the Americas, one species, the snowy egret, has been repeatedly mentioned as attractive to other species, perhaps because it is particularly good at finding patchy resources (Caldwell, 1981;

FIGURE 3.5 A mixed-species group of waders in a freshwater wetland in India. Egrets, gray heron, and black-headed ibis are photographed together in the Kolleru Lake reserve in Andhra Pradesh. Such groups of waders are probably more akin to an aggregation than to a group. *Picture from Wikimedia Commons, created by J.M. Garg.*

Master, 1992; Smith, 1995). White plumage, which the snowy egret has, has also generally been hypothesized to attract other birds (Kushlan, 1977; Caldwell, 1981), and social species are more white than nonsocial ones (Beauchamp and Heeb, 2001); however, experimental evidence that whiteness is per se more attractive than other colors is inconclusive (Green and Leberg, 2005).

Similar to waterfowl, waders have been noted to disturb ("beat") prey, by their movement through the water or by driving fish, which others then catch. Ibis (Courser and Dinsmore, 1975; Erwin, 1983), spoonbills (Russell, 1978; Kyle, 2006), and flamingos (Hurlbert et al., 1984) fall in this category. But these types of interactions can also occur when diving birds, such as cormorants or diving ducks or grebes, congregate fish into shallow areas. Waders appear to assemble in the shallows and wait for the diving species, because when the divers arrive, the waders catch fish with high frequency (Christman, 1957; Emlen and Ambrose, 1970; Bennett and Smithson, 2001).

3.5.5 Grassland Birds

Terborgh (1990) predicted that MSGs of grassland birds, similar to shorebirds, would be larger and of more fluid composition than forest MSGs, because in such environments with high visibility, predators attack prey by pursuit rather than ambush. Large group size is selected in these conditions for dilution of risk and a selfish-herd effect; the species may often look alike. Food is homogeneously distributed so that there is not a lot of niche partitioning among participants if food is rich; if it is not, the association terminates.

Empirical research on grassland MSGs is scarce, however. In general, they do appear to be loose groups, changing in their composition quite rapidly (Clergeau, 1990; Zamora et al., 1992). Some grassland systems do appear to mirror the predictions above in size and composition: for example, MSGs composed of the seed-eating sparrow and finch families can look quite similar to each other (Rubenstein et al., 1977), and such MSGs can be quite large (e.g., wintering flocks in the Californian desert averaged 50–200 birds; Cody, 1971). But other systems, especially those of ecosystems highly disturbed by humans, such as fields or lawns, can include diverse species adapted to such environments, including both granivores and insectivores (Barnard and Stephens, 1983; Clergeau, 1990; Veena and Lokesha, 1993). Most systems show seasonal changes in grouping propensity, with even tropical systems having less grouping during the breeding season, which is often linked to changes in rainfall (Tubelis, 2007).

Several papers on grassland bird MSGs stress the potential for high predation rates and the importance of vigilant species as leading species. Species may display sentinel behavior and make alarm calls (Alves and Cavalcanti, 1996; Ragusa-Netto, 2002), or rapidly detect threats (Greig-Smith, 1981). Indeed, the flight of other group members can be an alarm cue in birds of open areas; however, sometimes such a cue becomes unambiguous only when several birds have departed rapidly in succession (Lima, 1994; Proctor et al., 2001).

3.5.6 Forest Birds

Forest bird communities show the most complex MSG behavior of all animals. This phenomenon is found throughout the world on every continent other than Antarctica (Fig. 3.6), and the number of species per group is much higher than the other taxa we have discussed thus far, although the numbers of individuals per group is not necessarily large (Table 3.1). An extreme example of this diversity is demonstrated by groups in a Peruvian rainforest, where canopy groups included 53 participating species, and understory groups included 42 participating species (Munn and Terborgh, 1979; Munn, 1985). Sometimes these canopy and understory groups can combine, so the overall group contains 60–70 species and about 100 individuals (Munn, 1985). This environment is thus strikingly different from some of the other MSGs we have discussed thus far, in that for any one bird participating, the majority of other group members are heterospecifics, often with a wide range of characteristics. Because such a large number of species participate, the average propensity to be in a group ("prevalence" in Table 3.1, averaging across species) may be low, but this obscures the fact that there are usually several species for which the majority of records are in groups. Indeed, in some rainforests, more than 50% of individual birds of all species may be in MSGs at one time (e.g., Eguchi et al., 1993; Latta and Wunderle, 1996).

Which species are found in MSGs and which are not? These groups are not limited to birds: avian MSGs may even be joined by small species of squirrels, especially in tropical America (Della-Flora et al., 2013) and Asia (Kotagama

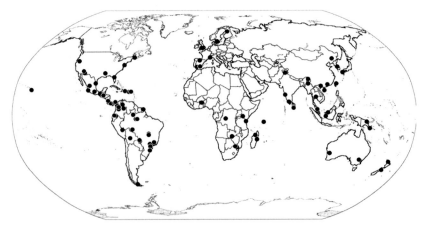

FIGURE 3.6 The geographical distribution of avian forest mixed-species group studies. Studies in the same place may be superimposed; sample size is 87 studies, including 93 sites that have published information on flock composition. *Adapted from Goodale, E., Ding, P., Liu, X., Martínez, A., Walters, M., Robinson, S.K., 2015. The structure of multi-species flocks and their role in the organization of forest bird communities, with special reference to China. Avian Research 6, 14 (as published under a Creative Commons license).*

and Goodale, 2004). Forest bird MSGs are usually composed of passerines birds, with some exceptions, including the woodpecker family, which are common following species throughout the world. In most parts of the world, especially temperate areas, MSGs are formed predominantly by insectivores (Powell, 1985; Greenberg, 2000). Some tropical MSGs, such as canopy groups in Amazonia that are dominated by tanagers, also include omnivores (Munn, 1985). Obligate frugivores such as parrots or barbets can sometimes enter into MSGs when fruit is evenly distributed, although their average propensity to be in a group is low (Kotagama and Goodale, 2004). Among insectivores, species that tend to group more are those that have a foraging ecology that makes them less vigilant for predators, either because they forage while looking at leaves or trunks of trees, or they forage in thick vegetation (Buskirk, 1976; Thiollay and Jullien, 1998; Thiollay, 1999, 2003). In addition, there appears to be an effect of predation intensity on the overall proportion of the avifauna that is in MSGs. MSGs are underdeveloped in islands with few predators (Willis, 1973b) and increase as raptor density increases (Thiollay, 1999).

Which species lead MSGs or are important to their formation and maintenance ("nuclear species" sensu Moynihan, 1962)? Leaders tend to be gregarious species (Goodale and Beauchamp, 2010) and breed cooperatively more often than would be expected by chance (Sridhar et al., 2009). Their sheer numbers and constant contact calling may make them easy to follow. Furthermore, because they may have related individuals or mates in the MSG, they may perform costly behaviors such as alarm calling, providing public information on which other species can then eavesdrop (Goodale et al., 2010; see Chapters 6 and 7). Leaders also tend to come from certain families. Throughout the northern temperate regions, MSGs are led by tits and, in some cases, have several species of tits per flock. Amazonian forest MSGs are led by antshrikes, with antwrens also playing an important role. Asian MSGs are led by gregarious babblers (usually inside forests) or white eyes (which may lead loose flocks in disturbed areas; Zhang et al., 2013; Mammides et al., 2015). These patterns indicate that to some extent the composition and structure of forest bird MSGs has a strong phylogenetic component, but we still await further studies to determine what percent of variation in these attributes is explainable by phylogeny (but see Gómez et al., 2010).

Like most forms of life and ecological interactions, forest bird MSGs have highest diversity in the tropics. Temperate MSGs may have considerably fewer species; for example, from the studies summarized in Table 3.1, temperate systems average 4.0 species for the midpoint of their range (n=4), whereas tropical ones average 8.1 species (n=4). MSGs in Central and South America are different from those in the rest of the world, in having only a few individuals per species; in Africa and Asia, some species, especially the leaders, are very gregarious (Goodale et al., 2009, 2015). Some tropical areas, especially those with high, stratified canopies, can have multiple group systems in the same place, with one system usually staying in the understory and the other in the midstory to canopy (Bell, 1983; Munn, 1985); shorter forests tend to have understory and canopy birds mixed together in groups (Poulsen, 1996). A few areas have been described where two or more MSG systems seem to coexist within ecosystems, with distinctions between large- and small-bodied groups (King and Rappole, 2001; Srinivasan et al., 2012; Mammides et al., 2015). In such places, birds tend to join MSGs with other species of their own body size (Mammides et al., 2015). This preference for similarly sized birds has been also seen through global meta-analyses (Sridhar et al., 2012), suggesting that competition (which usually requires species with disparate characteristics to coexist) is overwhelmed by facilitation (potentially through information sharing about predators and resources) in these systems.

Seasonally, the stability of these MSGs is largely dependent on latitude. Forest bird MSGs in temperate climates are mostly confined to the period after the young have fledged, the migratory season and winter. MSGs in tropical areas continue mostly throughout the year, and some continue even when their participating members and leading species are nesting (Munn, 1984; Jayarathna et al., 2013). In some tropical areas, winter migrants compose a majority of the members and can even lead MSGs (Hutto, 1987; Ewert and Askins, 1991), but in most areas the contribution of wintering birds to tropical MSGs is minor (Munn and Terborgh, 1979; Kotagama and Goodale, 2004). However, some tropical areas do show quite dramatic changes between wet and dry seasons (Davis, 1946; Develey and Peres, 2000).

The stability of forest bird MSGs also varies geographically. Compared to the ever-changing groups of grasslands, forest groups tend to be more stable, coming together in the morning and then dissolving only close to roosting time; however, they may become less obvious during the middle of the day, some-times bathing during this time (Jullien and Thiollay, 1998). One specific system in Amazonia is especially stable: the "core" species are not found outside groups and mutually defend an interspecific territory, hence staying together for much of their lives (Munn and Terborgh, 1979; Jullien and Thiollay, 1998). Indeed, the territories of these groups have been reported to remain constant for up to 17 years (Martínez and Gomez, 2013). In Central America, MSGs may also form around a small group of core species (Gradwohl and Greenberg, 1980). Elsewhere, the territoriality of forest bird MSGs is less clear, and the group may be more like a wave, with individual birds joining and leaving as it moves through their territories (McClure, 1967). More studies using banded birds are needed to really understand this.

The literature on forest bird MSGs is larger than that for any other taxa, including to our knowledge over 320 articles, and hence we cannot capture all of their complexity here. Nevertheless, we end with several general observations. These MSGs are one of the few mixed-group systems where joining has been shown to increase survival, at least for obligate members (Jullien and Clobert, 2000). Furthermore, dependencies between leaders and followers have been demonstrated experimentally, with the body condition of a following species reduced when a leader was removed (Dolby and Grubb, 1998). Finally, somewhat like the "protective mimicry" we discussed for fish, it has been postulated that birds in forest bird MSGs mimic each other in their plumage (Moynihan, 1968). Recent reassessment of such mimicry found some examples to be perhaps due to shared phylogeny or habitat, but other examples of species pairs in flocks do seem to exhibit striking similarities (Beauchamp and Goodale, 2011; Sazima, 2013; Section 6.3.3).

3.6 MSGS THAT INCLUDE MULTIPLE TAXA AND WHERE ONE SPECIES MAKES FOOD MORE ACCESSIBLE TO OTHERS

We now turn to MSGs with multiple taxa (i.e., members drawn from more than one of the taxa described earlier). We will not focus further on two such MSGs we have already mentioned in passing—mixed groups of crustaceans and fish (McFarland and Kotchian, 1982) and examples of squirrels interacting in bird flocks (Della-Flora et al., 2013)—in which animals of rather similar size or morphology interact. Rather, the phenomena in this section all consist of situations in which very different taxa interact, and these species-foraging techniques are compatible so that one species makes food more accessible to other species. But the types of MSGs discussed here differ in how symmetric and mutualistic the interaction is, with the first section—groups that follow "driving" or "beating" species—usually being a commensalism and the other two sections being true mutualisms.

3.6.1 Associations Based on One Species Increasing Prey Accessibility

Here we consider moving groups that center around, or follow, some animals that are not prey themselves but make food available to other species. When very different taxa move together in an MSG, odds are that one species provides food for the other through its movement. A famous example is given by cattle egrets that follow cattle or deer, consuming the insects that these ungulates disturb in the grass (Dinsmore, 1973; Fernandez et al., 2014). Studies demonstrate that birds capture prey at a rate higher than that when exploiting the same habitat in the absence of grazers, preferentially associate with active rather than resting grazers (Källander, 1993), and prefer some individuals of a herd more than others (Fernandez et al., 2014). Three complex systems analogous to this example have been described in detail: species that follow army ants, species that follow primates, and complex marine systems that include fish, aquatic mammals, and seabirds.

Army ants form large moving parties that sweep over everything in their path. All extant army ants apparently evolved from a single ancestor 100 mya (Brady, 2003). Two species of Neotropical ants make particularly large swarms that are followed by birds and even some lizards and frogs (Willis and Oniki,

1978; Willson, 2004), but birds on other continents, particularly Africa, are also known to follow army ants (Peters et al., 2008). The Neotropical family of antbirds include many species that are near-obligate ant followers (Brumfield et al., 2007), with some species showing specialized behaviors such as "bivouac checking," the inspection of ant nests in the morning to find active swarms (Swartz, 2001). The whole interaction appears to be somewhat parasitic on the ants, whose foraging rate is lowered in the presence of the associates (Wrege et al., 2005).

Another terrestrial beating relationship occurs between primates and other species that follow them. Primates (and their distant relatives the treeshrews, see Oommen and Shanker, 2009) disturb vegetation extensively, dropping both fruits and stirring up insects and small vertebrates and are commonly followed by many species of birds, including some predators such as raptors and mammals such as ungulates (Heymann and Hsia, 2015). On occasion, primates are followed by reptiles or even frugivorous fish, which trail behind primates that drop fruit into rainforest streams (Sabino and Sazima, 1999). The relationship is not purely that of beating but some following species can also benefit by responding to primate alarm calls (e.g., Rainey et al., 2004). Nonetheless, it seems rare that primates themselves benefit, so the relationship is mostly commensal (Heymann and Hsia, 2015).

In the oceans, an analogous phenomenon occurs with animals that drive fish. Yellow-finned tuna are predatory fish that drive their prey to the surface and are joined by species of dolphins and seabirds (Au and Pitman, 1986). This phenomenon is complicated: in some areas the tuna are not necessary and the driving of fish is done by diving birds (Anderwald et al., 2011), dolphins (Vaughn et al., 2007), or even larger whales. This driving aggregates the fish into a "ball" that makes it easier for other species to attack en masse (Grover and Olla, 1983; Clua and Grosvalet, 2001; Vaughn et al., 2007). Surface-feeding or shallow-diving species attack the ball as it gets closer to the surface (Hoffman et al., 1981), and a few species can be completely dependent on the interaction (Pitman and Ballance, 1992). Predatory fish such as sharks may also follow tuna groups, both as predators and as scavengers on leftovers, so that a whole community moves together through the seas (Au, 1991). Yet, interestingly, some recent research on these groups suggests that one of the main benefits for tuna and dolphins in associating is that they collectively decrease their predation risk (Scott et al., 2012).

The phenomena above are by and large commensal, with the leader species being unaffected (or slightly parasitized in the case of the army ants), which contrasts to most of the other mutualistic MSGs discussed in this chapter. Nevertheless, driving and beating also occur within other MSG systems, affecting some species more than others. For example, in bird MSGs, there are some sallying birds that specialize in foraging on insects that are disturbed by leaf-gleaning species (Munn, 1984; Hino, 1998; Satischandra et al., 2007; Sridhar and Shanker, 2014b).

3.6.2 Cooperative Hunting

Cooperative hunting between distantly related taxa provides another example of MSGs in which one species makes prey more accessible to other participants. Take, the case of coyotes and badgers hunting ground squirrels (Minta et al., 1992). Badgers pursue the squirrels under the ground whereas coyotes wait for the squirrels at the burrow entrance, blocking their escape route, and combined hunting success is increased when the squirrels are attacked in these two ways. In Section 4.2.5, we describe a very similar aquatic interaction of two species with different but compatible hunting techniques in detail—eels and groupers—cooperating in coral reefs. Another example in mammals may occur between jackals and cheetahs. Eaton (1969) observed jackal running into herds of ungulates and barking in a way the author interpreted as distracting the prey, which were then successfully predated by cheetah. Successful kills by cheetah generated carcasses that jackals scavenged after cheetah had had their fill. This interesting behavior appears not to be widespread, as no follow-up study seems to have been done.

Other examples of cooperative hunting involve humans. In Section 4.2.5, we describe in detail the complicated relationship between honeyguides and humans, who share the goal of finding and obtaining nourishment from beehives. Humans and dolphins may also use each other to find food (Scott et al., 2012; Ashley Ward, personal communication). Some research suggests that analogous communication may have influenced the hunting between humans and the ancestors of dogs, driving dog domestication (Miklósi, 2009). Of course, we should mention in passing that humans and domestic animals represent a recently evolved form of MSG, with large cultural and economic driving factors that make them beyond the scope of this book.

3.6.3 Mutualisms in Which Increased Foraging Is Traded for Vigilance

In some relationships between mammals and birds, food can be exchanged for vigilance. The first report of this kind of behavior was by Rasa (1983), who described a mutualism between mongooses and hornbills, in which the hornbills acted as sentinels for the mongooses but also increased their foraging in association with them, as the mongooses' digging and movements expose insects that would ordinarily not be accessible. The phenomenon is altruistic to the degree that the hornbills make alarm calls to predators that are not threats to themselves but are to mongooses (Rasa, 1983; Section 5.4.2).

Since this study, similar MSG systems have been described between mongooses and other birds. For example, dwarf mongooses also associate with drongos, which provide alarms and allow the mongooses to reduce their vigilance (Sharpe et al., 2010). Drongos, however, can also be manipulative and sometimes steal food from the mongooses they associate with through deceptive

FIGURE 3.7 Drongos perform sentinel behavior when associated with mongooses but also manipulate their behavior. Drongos have been reported to make false alarm calls and also mimic the alarm sounds of other species, both when predators are present and when they are not (Flower, 2011; Flower et al., 2014). *Photography by Tom Flower, used by his permission.*

alarms, sometimes including vocal mimicry (Flower, 2011; Flower et al., 2014; Fig. 3.7), and we will look again at these kind of associations in a case study in Chapter 5 (Section 5.4.5) and in discussing communication within MSGs in Chapter 6. Mongooses can also share vigilance with other mammals that nest near them, such as Cape ground squirrels (Makenbach et al., 2013). The MSGs described here involve three different species of mongoose; mongooses may be preadapted to participate in such interspecific interactions because of the importance of sentinel behavior in their own monospecific societies (e.g., Kern and Radford, 2013).

3.7 CONCLUSIONS

In reviewing the diverse habitats and groups of taxa discussed in this chapter, we hope that commonalities have become apparent. For example, think of lone individuals of social species that join MSGs to be part of a group. We have encountered these in as diverse groups as mysids (Wittmann, 1977), dolphins (Frantzis and Herzing, 2002), and primates (Fleury and Gautier-Hion, 1997) or one species that associates with a more vulnerable species, hence reducing its predation risk, suggested to occur in fish (Mathis and Chivers, 2003) and ungulates (FitzGibbon, 1990). Species that make prey available to other species have been seen to play special roles in groups in this chapter, such as in coral fish (Sazima et al., 2007), wading birds (Russell, 1978), and waterfowl (Källander, 2005), as well as the driving and beating examples in Section 3.6.1. Vigilant species playing important roles can be found across many taxa, such as in Diana monkeys (Noë and Bshary, 1997), shorebirds (Gavrilov, 2015), and grassland birds (Greig-Smith, 1981).

Table 3.1 has so much variation within a taxon that differences between taxa are largely obscured, other than the higher species diversity in bird MSGs, particularly pronounced in forests. Perhaps this within-taxon variation is the most important story to be gleaned from this data. Of course, we must remember the importance of definitions in this exercise. Compare 2% prevalence of MSGs in waterbirds when a group is defined as a bird that lands within 2 s of another (Silverman et al., 2001) to 90.1% prevalence in Scandinavian lakes, when a group is defined as a "collection of ducks foraging in a bay" (Pöysä, 1986b). Yet at the same time, surely some of this variation is because of environmental conditions, such as resource availability, breeding seasonality, habitat density, and many other factors, and it is stimulating to think how much variation there is to explain. Studies that look at variation of the prevalence of MSGs in relationship to predation (e.g., Thiollay, 1999; Chapman and Chapman, 2000) are particularly valuable because of the importance of predation to explanations of MSGs (Chapter 5). Given the widespread distribution of MSG across taxa described here, a particularly exciting direction might be to relate MSG prevalence to predation risk for different taxa, using a standard measurement of that risk, such as attacks per individual per time interval when foraging on their own.

Chapter 4

Adaptive Implications of Mixed-Species Grouping: Foraging, Physical, and Reproductive Factors

In this chapter and the next, we consider the costs and benefits of mixed-species grouping in animals in detail (and in plants only briefly). The bulk of the chapter is devoted to considering the foraging costs and benefits of being in a mixed-species group (MSG), before considering consequences for the energetics of loco-motion, protection from the elements, and reproduction. In the next chapter our primary focus will be on protection from predators and parasites, before we finish with a number of case studies that highlight the complex interplay between different costs and benefits.

In this chapter and the next, we draw examples very widely to explore a broad range of relevant mechanisms. Sometimes, we use cases that are more strictly mixed-species associations (MSAs) rather than MSGs and even some cases where individuals of different species simply find themselves in close proximity. Nevertheless, we are most interested in self-organized groups, where the stimulus to join a group is the group itself and not an external factor, and we will mostly concentrate on MSG and use that acronym. That is particularly true for this chapter, as for many aggregative MSAs the foraging benefit is simply finding the food resource they are massing around.

4.1 DIFFERENT TYPES OF MIXED-SPECIES GROUPS IN TERMS OF ADAPTATION

In Chapter 1, we defined MSGs very much in empirical terms. We chose not to use a definition in terms of the selection pressures acting, because in many systems these have not been fully characterized. Despite this, we would expect that being in close proximity to other species will impose costs and benefits on the individuals concerned, and in this chapter and the next we will explore these. It is well known that being in a group can have a wide variety of consequences for the individuals involved (Giraldeau and Caraco, 2000; Krause and Ruxton,

2002; Sumpter, 2010; Beauchamp, 2014); here we will explore these particularly within the context of mixed-species grouping.

Let us consider the fitness of an individual of some species A as a consequence of it being in a group comprised of m individuals of its own species and n individuals of another species B. We denote this fitness $f_A(m,n)$. A number of possibilities exist.

Firstly, it could be that the fitness of an individual of species A is higher in that of MSG of size $m+n$ than it would be if it was in a group of $m+n$ individuals of its own species (i.e., $f_A(m,n) > f_A(m+n,0)$). This indicates that there is some special benefit (what Beauchamp (2014) calls a "unique benefit") to individuals of species A arising from grouping with individuals of species B. Putting this in a different way, the fitness of individuals of species A is affected not just by group size but by group composition, and associating with individuals of species B confers special benefits. Such a situation could lead to observation of a behavioral preference shown by individuals of species A for heterospecifics over a similar-sized group of conspecifics. Another possible situation is that an individual of species A has exactly the same fitness in the MSG as it would in a group of $m+n$ individuals of its own species ($f_A(m,n) = f_A(m+n,0)$). In this case, individuals of species A are unaffected by the composition of the group they are in and are affected by group participants of its own species and species B identically. There are no special benefits or special costs to an MSG in this situation, and we might expect to observe no behavioral preference for heterospecifics over conspecifics or vice versa.

The next situation is that an individual of species A in an MSG has a higher fitness than if it were in a group of only m individuals of its own species but less than if it were in a group of $m+n$ of its own species ($f_A(m+n,0) > f_A(m,n) > f_A(m,0)$). This implies that there is a benefit to being in a larger group, but this benefit is reduced if larger size can only be achieved by association with the other species. This situation might translate into observation of a preference for joining a group of conspecifics over a similar-sized group of heterospecifics or a similar-sized MSG.

It could be that an individual of species A in the MSG has exactly the same fitness as if it were in a group of m conspecifics ($f_A(m,n) = f_A(m,0)$). In this situation, association with individuals of species B has no positive or negative overall effect on individuals of species A. In this situation we might expect behavioral indifference to species B, with neither attraction nor avoidance shown.

Finally, there could be a situation where an individual in the MSG has lower fitness than an individual in a group of m conspecifics ($f_A(m,n) < f_A(m,0)$). In this case, association with species B individuals has an overall negative effect on species A individuals, and we would expect individuals of species A to avoid an association with species B individuals. See Table 4.1 for a summary of these five conditions.

Of course, in the MSG described above, there is also potential for A individuals to impact on B individuals. A situation where individuals of both species

TABLE 4.1 Different Net Fitness Consequences of Being in a Mixed-Species Group

Biological Circumstance	Fitness Consequence	Behavior
B individuals offer a particular benefit	$f_A(m,n) > f_A(m+n,0)$	A individuals actively prefer mixed-species groups
B individuals offer identical benefits to A individuals	$f_A(m,n) = f_A(m+n,0)$	A individuals show no preference in terms of group composition
B Individuals offer a benefit but less than A individuals	$f_A(m+n,0) > f_A(m,n) > f_A(m,0)$	A individuals will join mixed-species groups but prefer same-size groups of their own species
B individuals have no effect at all on A individuals	$f_A(m,n) = f_A(m,0)$	A individuals are indifferent to the presence of B individuals in a group
B individuals impose a cost on A individuals	$f_A(m,n) < f_A(m,0)$	A individuals actively avoid grouping with B individuals

We delineate five different biological circumstances, in terms of the fitness of an individual of species A of grouping either with its own species or with a mix of that species and another (species B). We use $f_A(m,n)$ to denote the fitness of an A individual in a group of m A individuals and n B individuals. We also suggest the likely group selection preference of A individuals in each circumstance.

benefit from the MSG in comparison to the two single-species subgroups could be termed a *mutualistic MSG*. In this situation, there is no reason why the benefits need be equivalent for both parties, nor why they should be expressed through the same mechanism. It might be that one species experiences a modest reduction in predation risk and the other experiences a substantial improvement in foraging rate. Similarly, there is no reason why the overall benefit cannot involve the summation of several different effects; for example, a species might experience a net fitness benefit from participation in an MSG because an antipredatory benefit outweighs a cost experienced through increased competition for food.

A situation where one species gains a benefit from the MSG but the other is entirely unaffected by the presence of heterospecifics could be called a *commensal MSG*. A situation where one species benefits but the other experiences a fitness cost of an MSG (relative to the group fracturing into two single-species subgroups) might be called an *exploitative MSG*. Such situations might occur where it is impossible for individuals of the exploited species to force fracturing

TABLE 4.2 Different Types of Mixed-Species Groups in Terms of Costs and Benefits

Fitness Situation	Description
Both species benefit from being in the mixed-species association	Mutualistic MSG
One species benefits but the other is entirely unaffected	Commensal MSG
One species benefits but the other suffers	Exploitative MSG
Both species suffer from being in an MSG	Mutually disadvantageous MSG

We consider two species, and compare their fitness when in a group together versus when that group is split into two monospecific groups. *MSG*, mixed-species group.

of the MSG into single-species subgroups, or the cost of forcing such a fracture is greater than the cost of remaining in the MSG. Such exploitation need not necessarily take the form of parasitism where the exploiter uses the victim as a resource. It could (for example) occur in a foraging context where one species can steal food discoveries made by the other.

It might even be possible to observe situations where MSGs occur even though both species would benefit from a split of the group into single-species subgroups (*mutually disadvantageous MSGs*). This might occur if recent environmental change has increased the costs of the MSG to one or both species, and the species concerned do not have the phenotypic flexibility and/or have not experienced sufficient genotypic change to allow removal of the tendency to form maladaptive MSGs. These situations have been summarized in Table 4.2.

In the rest of this chapter and the next, we briefly consider a range of mechanisms that might lead to benefits or costs of being in a group, with a particular focus on MSGs. Associating with heterospecifics can sometimes provide benefits associated with being in larger group in situations where conspecifics are not readily available (examples in the last chapter included mysids, dolphins, and primates; Section 3.7). We will not dwell on consequences that stem from an increase in *group size* as a result of the formation of an MSG but rather focus on consequences of *group composition*.

4.2 SOME POTENTIAL FORAGING BENEFITS OF (MIXED-SPECIES) GROUPING

We separate out foraging-related benefits in this chapter and antipredator benefits in the next chapter; however, foraging and guarding against predation are often strongly interrelated activities. As an example, there are situations in primate

MSGs where one species will exploit food in some parts of the environment more willingly when associated with another species that offers a high ability to warn of impending predator attacks (Porter, 2001; McGraw and Bshary, 2002). Similarly, associating with a more vigilant species might allow individuals of another species to decrease their personal investment in antipredatory vigilance and thus devote more time or simply more attention to food finding. Wolters and Zuberbühler (2003) observed that MSGs of Diana and Campbell's monkeys both exploited more of their environment while foraging, and foraged more intensively, than when in single-species groups. The authors argue that this is likely because the complementary vigilance patterns of the two species increase the ability to detect predators. Despite such interactions, we still consider that there is practical value in focusing on the specific mechanisms by which foraging could be enhanced in an MSA.

4.2.1 Sharing Information About Foraging Opportunities

Discovering food through detection of the food discoveries of others (a process called *local enhancement*) seems widespread in the animal kingdom (Galef and Giraldeau, 2001). The conditions for it to be effective are that: (1) food is scattered throughout the environment and challenging to find; (2) another forager or foragers in the process of exploiting discovered food is more easily detectable than undiscovered food; and (3) a discovered food item cannot be entirely monopolized by the individual that first discovers it. When these conditions are met, it can sometimes be more economic for a forager to break off from searching for as yet undiscovered food itself and respond to the food discovery of another nearby individual by attempting to share or usurp that discovery (Beauchamp and Ruxton, 2014). Since proximity is a key aspect, local enhancement is associated with aggregation. In general, the benefits to an individual of local enhancement will increase with increasing group size (at least at small group sizes), and so there may be a selective benefit to MSGs between species that share the same food to reach the association size that brings optimal net benefits from local enhancement.

The costs and benefits of local enhancement are frequency dependent (Barnard and Sibly, 1981). In a group of individuals that ignore the activities of others, an individual that also uses local enhancement could increase its net rate of food consumption. This effect might lead to the spread of the strategy within the group. However, the more group members engage in this activity, the more the net rate of food discovery by the group as a whole declines (if responding to discoveries of others takes a finite time and as yet undiscovered food is less likely to be detected by the responder during this time). The net rate of food discovery by the group as a whole might even decline when these conditions are not met, if the process of local enhancement leads to overlap of search areas of individuals that recently shared a discovered food patch until it was exhausted (Beauchamp, 2008). Even when all individuals are exploiting local enhancement and the aggregate rate of food discovery by the group is decreased, it may still pay individuals to continue with this strategy because an individual that unilaterally gave up the benefits of local enhancement would still pay the costs of others sharing or usurping its discoveries. Local enhancement then may actually lead to reduced average individual rate of food gain relative to a similar set of individuals that are somehow prevented from using it, but local enhancement will still be an evolutionarily stable strategy from that perspective. Under all circumstances, local enhancement should reduce the variance in time between food patch encounters and may be selected for that reason.

Both the costs and benefits to an individual will vary with the number of foragers involved. Specifically, as group size increases, opportunities to take advantage of food discoveries of others will increase, but the value of each discovery to both the original discoverer and any subsequent joiners will decline through increased competition. The optimal group size from an individual's perspective will be affected by the ease with which food patches can be found, the value of those patches, the ease with which exploited patches can be detected by other foragers, the speed with which exploited patches can be reached, the nature of competition between individuals on a patch, and the trade-off (if any) between searching for undiscovered and discovered patches (Giraldeau and Caraco, 2000). However, sometimes an individual will benefit from joining a group, and this effect of group size may be sufficient to create MSAs/MSGs via local enhancement.

Individuals can also obtain additional benefits from associating with particular heterospecifics over and above the benefits they would gain by increasing group size alone. For example, it can be more beneficial to forage with a species that is good at finding food, that is easier to detect when exploiting food, and/or which can be outcompeted at a food sources. This is clearly the case in a community of small passerine birds in wooded areas of England. One species in particular, the marsh tit, is more likely to uncover resources than other local species. When mapping the spread of information about novel food sources through mixed-species social networks (Fig. 4.1), they were shown to provide a disproportionate amount of information about the location of resources in their habitat (Farine et al., 2015a). This information was used by conspecifics and heterospecifics alike and served to build MSGs. In this system, we would thus expect other species to pay particular attention to marsh tits when choosing feeding locations. The process of information spread could also be actively promoted by particular species. For example, willow tits (closely related to marsh tits) in Japan were observed making recruitment calls when they discover new food sources that both conspecifics and individuals of other species responded to, facilitating formation of MSGs (Suzuki, 2012). Suzuki speculates that the willow tits call to gain antipredatory benefits of flocking even though there may be competition costs.

Vultures are the iconic examples of local enhancement. They often circle in the air for some time above a potential food source before landing to

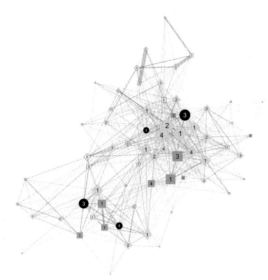

FIGURE 4.1   An interaction network for three tit species. The figure has nodes for individuals, and lines between nodes represent propensity to forage together. Here the marsh tits are in *black circles*, great tits are *squares* (dark gray or white), and blue tits are *diamonds* (light gray or white). White identifies birds that have never discovered food, size is the number of connections a node has, and numbers are the number of trials in which each bird discovered food (out of 4). From this figure, it is clear that birds that discovered food were more central, while birds that never discovered food (those with "0") are typically less well connected (smaller nodes). Three of the four marsh tits discovered at least three food sources (they are disproportionately represented in that sample), and in doing so, introduced much more information into the network than the other individuals did on average. *Adapted from Farine, D.R., Aplin, L.M., Sheldon, B.C., Hoppitt, W., 2015a. Interspecific social networks promote information transmission in wild songbirds. Proceedings of the Royal Society of London B: Biological Sciences 282, 20142804 and provided by Damien Farine.*

exploit it. This behavior makes them ideal for local enhancement, and so both other flying vultures and terrestrial scavengers respond to circling vultures (Morelli et al., 2015). However, it has been demonstrated that vultures themselves benefit from association with smaller raptorial species through local enhancement in Kenyan grasslands. Kane et al. (2014) demonstrated that vultures obtained information on carcass location from scavenging eagles. Here the key appears to be that eagles have superior ability to find prey. Both the eagles and the vultures wait until there are thermals of rising warm air that they can use for soaring flight before beginning their foraging activity each day. But eagles are smaller and lighter and so can take advantage of weaker thermals forming earlier in the day. Thus by the time the thermals become strong enough to support the vultures, the eagles have already been in the air for some time and have discovered animals that died or were predated overnight nearby; the vultures then follow the eagles to their discoveries (Fig. 4.2).

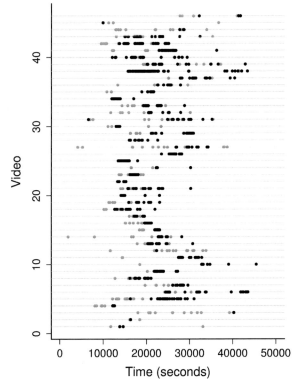

FIGURE 4.2 **Arrivals of eagles and vultures to scavenge from a carcass.** Graph of the recorded arrival times of vultures (*black points*) and scavenging eagles (*gray points*) to 46 experimental carcasses set out in the Laikipia district of Kenya. Eagles arrived before vultures in 38 out of 46 cases and were close in time in a significant number of cases to suggest that eagles were being followed by vultures. *Adapted from Kane, A., Jackson, A.L., Ogada, D.L., Monadjem, A., McNally, L., 2014. Vultures acquire information on carcass location from scavenging eagles. Proceedings of the Royal Society of London B: Biological Sciences 281 and provided by Adam Kane.*

In another example, South African Augrabies flat lizards feed on ripe figs and use the activity of frugivorous birds as a means of detecting trees with ripe fruit from a distance. Here the lizards benefit from the greater ability of the birds to search widely and quickly in the local environment. Whiting and Greeff (1999) demonstrated the importance of bird activity in attracting lizards to trees in a manipulative field experiment. Peres (1996) reported on the consequence of food searching on naturally occurring MSGs of saddle-backed and red-capped moustached tamarins in Amazonian forest. The former searched lower in the canopy and found small patches relatively frequently; the latter searched higher where there were less frequent but more valuable patches. Each species alerted the other to finds, even though (as we dis-cuss later) the moustached tamarins were able to competitively exclude the saddlebacks from small patches after they arrived at them. Insectivorous bats have repeatedly been demonstrated to be sensitive to the echolocation calls of sympatric heterospecific bats, being particularly drawn to those of species that have the strongest dietary overlap (Li et al., 2014).

The cross-species information transfer required for local enhancement has been demonstrated unequivocally in the laboratory in experiments that uncovered information transfer not just about patch location but even also patch quality (Coolen et al., 2003; Webster et al., 2008; Boulay et al., 2016). Local enhancement has long been considered important to foraging seabirds (Haney et al., 1992), and this has been demonstrated experimentally in the field using seabird-mimicking plastic models placed on the ocean surface (Bairos-Novak et al., 2015). Although seabirds in this experiment responded most strongly to models of their own species, they also showed cross-species attraction. These studies highlight a potential limit to the use of local enhancement as a force to form MSGs. Within a species, individuals exploit the same foraging niche. However, different species are expected to exploit different niches especially in sympatric areas. We would expect local enhancement to be more prevalent in species that have a greater overlap in their niches and that use similar signals to attract others.

A persistent interest in the group foraging literature has been the information center hypothesis, the idea that areas where foragers come to rest or breed can provide opportunities to gain information about distant foods (Ward and Zahavi, 1973). Despite much discussion, this mechanism seems much less common in a within-species context than local enhancement (Richner and Heeb, 1995), with only a few well-documented cases (e.g., Marzluff et al., 1996; Sonerud et al., 2001). Furthermore, information sharing between species has never been reported, perhaps because individuals of different species tend to leave the colonies at different times (Erwin, 1983). For example, Weimerskirch et al. (2010) studied the Guanay cormorant and Peruvian booby that breed together and feed on the same prey. Although the cormorants used social information from their own species to select the direction of their departure from the colony, they did not use information from boobies, nor were the boobies influenced by the cormorants.

4.2.2 Beater Effects and Enhancement of Food Availability

One species can forage in such a way that makes food that was previously inaccessible or difficult to find, more available to members of another species. In the last chapter, we discussed several examples of beating and driving that have been studied in detail, including followers of army ants, groups that form around primates, and mixed associations of tuna, dolphins, and seabirds (Section 3.6.1). We also discussed mongoose associations that provide more foraging opportunities to birds (Section 3.6.3); refer Section 4.2.4 about cooperative hunting, which centers around species whose foraging techniques are compatible. These

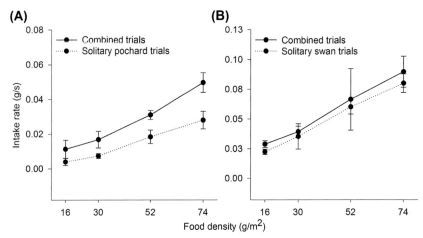

(A)

(B)

FIGURE 4.3 **The apparent functions of associations between primates and nonprimates.** In primate–nonprimate associations, benefits are asymmetrically distributed between primates and nonprimates, with nonprimates being the beneficiary much more frequently than primates. This is independent of the type of benefit (foraging or antipredator) and the geographic region. In the majority of cases, foraging benefits include feeding on prey flushed or on residuals (leaves, fruits, seeds) dropped by primates. *Adapted from Heymann, E.W., Hsia, S.S., 2015. Unlike fellows – a review of primate–non-primate associations. Biological Reviews 90, 142–156 and provided by Eckhard W. Heymann.*

types of associations can range from those that are parasitic on the beaters (e.g., army ants), to mutualistic associations (e.g., the mongooses). However, interactions in which the beaters are not much affected by the presence of followers are the most usual situations, as exemplified by an analysis of which species are benefitted in mixed taxa MSGs that follow primates (Heymann and Hsia, 2015, Fig. 4.3)

Here we concentrate on a few examples of beating and driving that demonstrate the influence the beaters have on communities, both those moving and those stationary, those composed of plant eaters, and also those composed of animals that eat other animals. We focus first on swans, which, as mentioned in Section 3.5.3, are followed by many other waterfowl species because they disturb vegetation by virtue of their large size and long necks. In particular, Bewick's swans often forage by trampling in the water to excavate tubers from the substrate. This trampling releases small animals and small parts of vegetable matter into the water column that are consumed (after it resettles on the sediment surface) by diving and dabbling ducks that preferentially associate with Bewick's swans over other swans (such as mute swans) that do not excavate with their feet. Gyimesi et al. (2012) demonstrated in an experimental setting that this association allowed enhanced feeding by the ducks at no cost to the swans (likely because the excavated material is too dispersed to be profitable for the larger swans to gather but can be harvested efficiently by the smaller ducks, Fig. 4.4). Interspecific aggression among the

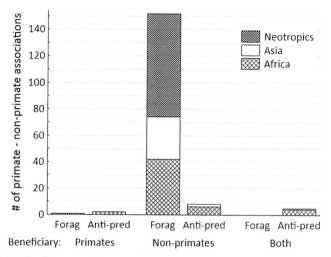

FIGURE 4.4 Feeding rates of two water-bird species, alone and together. Calculated instantaneous intake rates of pochards (left) and swans (right) when foraging alone ("solitary trials") or accompanied by the other species ("combined trials"). Given are means±standard errors. Note the different scale ranges on the y-axes. *Adapted from Gyimesi, A., van Lith, B., Nolet, B.A., 2012. Commensal foraging with Bewick's Swans* Cygnus bewickii *doubles instantaneous intake rate of Common Pochards* Aythya ferina. *Ardea 100, 55–62 and provided by Abel Gyimesi.*

waterfowl followers is often reported, with species jockeying for the best positions behind the swans (Bailey and Batt, 1974).

Our stationary example of animals making food more accessible for other species is the case of a solitary orb-weaving spider increasing its foraging when its web was attached to those of a colonial orb-weaver species (Hodge and Uetz, 1996). The spider's rate of food uptake was greater in these conditions than when alone or attached to the webs of members of its own species, and this led to a measureable benefit in fecundity. The result seems to be explained by the webs of the colonial species allowing construction of webs higher off the ground, where more food was available. In addition, food that was not caught by one net can bounce on the nets of other spiders (the "ricochet effect," Uetz, 1989), which reduces the number of prey that can escape.

4.2.3 Niche Separation, Reduced Competition, and Altered Social Interaction

It may seem obvious that individuals of the same species will have similar or identical foraging niches and so will suffer high levels of competition when sharing the same food patch; an MSG might allow a reduction in such competition because different species will have less similar foraging niches. Thus, this line of reasoning goes, an MSG can allow species to gain some advantages of grouping (say in reduced risk of predation) while paying lower costs (relative to a single-species group) in terms of competition for food. This reasoning is logically solid, but there are some important complications.

Firstly, there can be a very strong overlap in the foraging niches of different species. Graves and Gotelli (1993) used this to explain their observation of MSGs involving three woodcreeper species in Peruvian rainforest. Two similar-sized species never cooccurred in the same group, despite both of them commonly cooccurring separately with the third larger species.

Secondly, competition costs may be reduced in MSGs, but they may still be nontrivial. Sridhar et al. (2012) cite a number of examples of niche divergence in MSGs. That is, two species shift their ranges of resources exploited in different directions compared to when foraging without the other, so as to reduce overlap in these ranges and thus reduce competition. This niche shifting almost certainly means foregoing otherwise-attractive food-types and/or accepting otherwise unattractive foods to reduce competition with group mates.

Furthermore, some competition may be inevitable if group members are to have sufficient commonality of interest in how they use the environment to keep the MSG together. That is, if the benefits accruing from an MSG are to be used, then some resource competition may be the inevitable price that must be paid for this, and some of these benefits may relate to foraging. For example, in local enhancement, it is only attractive for an individual of one species to be drawn to feeding individuals of those species that have at least some commonality of diet (Seppänen et al., 2007). In keeping with this, Sridhar et al. (2012) found a positive relationship between the association strength and phenotypic similarity of species in avian MSGs worldwide. Thus, while some difference in foraging niche between species may bring an advantage to MSGs in terms of reduced competition, this effect will be tempered by a need for shared diet to maintain motivation of the individuals concerned to remain in the MSG.

Finally, competition may be attractive to some members of an MSG because they can obtain a benefit through exploiting members of another species (e.g., by attacking and usurping members of a subdominant species when they discover food items). In some situations, competition is costly to all parties. For example, for an MSG of birds feeding on worms on intertidal mudflats, interference can take the form of one bird simply getting in the way of another or worms responding to the close passage of one bird by burying themselves deeper, thus making themselves less available to subsequent birds. These forms of competition will, in general, lead to costs to all members of a group. However, if one of the species is more aggressive and socially dominant to the other, then there may be another form of competition where a worm found by a member of the competitively inferior species is sometimes stolen by a member of the other species before it is swallowed by its initial discoverer. This kleptoparasitic interference competition is clearly costly to the competitively inferior species but is an advantage to the other species that accrues from being in an MSG (Barnard and Thompson, 1982).

A good example of the complexity of competition in MSGs is a study of the Madagascar paradise flycatcher and common newtonia that frequently form two-species flocks in the deciduous dry forest of western Madagascar (Hino, 2000). In the flycatcher, some males have long tails, whereas other males and all females have short tails. When foraging in MSGs, each type of bird in this species captured its prey more rapidly than otherwise (likely because of a beater effect of the activity of the newtonia). However, the degree of increase in feeding rate was smaller in long-tailed males. When in an MSG, all flycatchers caught prey on leaves in the canopy where newtonia foraged. When not in an MSG, long-tailed males caught insects in midair by sallying, whereas the other types again gleaned from leaves. The elongated tails of long-tailed males may hamper the agility required for gleaning from leaves, and this may be a foraging cost to the sexually selected long tails that bring a benefit in increased attraction of mates. Hence, the foraging shift in MSGs favored one morph of this species more than the other.

The other member of this association, the common newtonia, is monomorphic and often formed groups of three to five individuals. In monospecific flocks, subordinates fed at low rates on branches owing to frequent aggression from dominants when attempting to feed from leaves. When foraging in an MSG, however, subordinates foraged among leaves, and their feeding rates increased because the frequency of intraspecific interference decreased greatly (likely because flycatcher individuals were often physically interposed between newtonia individuals). Dominants did not show any difference in feeding pattern with social situation but are thought by the study author to gain antipredatory benefits from grouping.

Experimental manipulations are relatively uncommon in the study of MSGs but the study of Sasvári (1992) on the foraging of MSGs of great tits and marsh tits is an exception. Sasvári provided artificial food patches on which birds had to search for hidden prey items. Great tits showed increased foraging rates when feeding in MSGs compared with same-sized single-species groups. Sasvári put this effect down to high levels of aggression between great tits that were dampened by marsh tits, often being interposed between them. Female great tits are subdominant to males and benefit particularly from an MSG. Marsh tits were subdominant to all great tits and fed at a lower rate than them in MSGs as a result of a considerable fraction of time spent avoiding aggression from great tits. Sasvári argues that marsh tits form MSGs rather than feeding alone or in single-species groups because great tits are common relative to both marsh tits and feeding patches, and the availability of food patches that can be exploited by a marsh tit without attracting great tits will be low in natural circumstances. This was especially true in the experiment because great tits were preferentially drawn to a patch with marsh tits already on it, rather than one with a similar number of conspecifics. Sasvári also varied the food density on the patch and found that when food (and so the exploiting group of birds) was more spread out, then these reduced aggressive interactions caused an increase in the foraging rate of all individuals, especially the marsh tits at the bottom of the dominance hierarchy.

Similarly, Peres (1996) reported on the consequence of food patch size for naturally occurring MSGs of saddlebacked tamarins and red-capped moustached tamarins in Amazonian forest. When patches of fruit were densely packed, the dominant moustached tamarins excluded the other species, but this was not possible on larger trees with more dispersed fruits.

Perhaps the clearest example of the potential for complexity of competition effects in MSGs comes from the meticulous observations of Barnard and Thompson (1982) on groups of birds feeding on estuarine worms. These groups generally contained both lapwings and golden plovers, sometimes joined by black-headed gulls that did not discover worms themselves but rather chased members of the other two species, inducing them to drop recently captured worms.

In flocks without gulls, lapwing net rate of energy intake increased with the number of conspecifics but was unaffected by the number of plovers; plover intake, in contrast, was negatively affected by the number of lapwings but unaffected by the number of conspecifics. In the presence of gulls, however, lapwing intake was no longer influenced by conspecific number but declined with increasing number of plovers; in plovers, intake rate now increased with the number of conspecifics but was unaffected by lapwings. Overall, the presence of gulls was associated with a decrease in foraging efficiency in lapwings but actually led to an increase in plovers.

This complexity of responses has a number of underlying drivers. Firstly, gulls joined flocks nonrandomly with respect to flock size, the ratio of plovers to lapwings in the flock, and the type of worms being taken. Furthermore, lapwings and plovers responded to the arrival of gulls in a number of ways, including departure from the group by lapwings, change in the positioning of individuals of both species within the group, and a switch to smaller worms that were less attractive to kleptoparasitic gulls.

In groups without gulls, lapwings will steal food from nearby plovers as well as finding their own, and Barnard et al. argued that plovers join lapwings to take advantage of their superior antipredatory vigilance (in 93% of flushes, it was a lapwing that took off first). However, gulls concentrated their food stealing on lapwings that released worms more readily than plovers did. When the flock is dominated by plovers, spatial clumping causes a reduction in lapwing foraging efficiency, and this is more likely to occur after lapwings leave the flock in response to gull arrival. Lastly, when gulls arrived, those lapwings that remained tended to clump together more, which reduced not only their exposure to gulls but also the potential to steal worms from plovers.

4.2.4 Reduced Variance in Food Finding

As discussed earlier, being in a large group and using local enhancement should lead to reduced variance in intervals between food patch encounters, and this is true in both single-species groups and MSGs. Although this benefit to grouping is uncontroversial, we know of no instances where this is considered to be the primary driver for aggregation. Nor do we expect that this mechanism is likely to lead to selection for MSGs over single-species groups for reasons over and above those discussed earlier under local enhancement. However, this benefit has been observed in MSGs. For example, Podolsky (1990) described how MSGs of primates in Peruvian forest were able to reduce the variance in times spent searching for a new fruiting tree to feed from, because a larger group had a broader search area and crucially because different species had different characteristic search paths. Certainly, more research on this mechanism is warranted.

4.2.5 Group Hunting

Hunting of live prey by groups of animals is widespread (Bailey et al., 2013). This can offer a number of benefits relative to predation by a single individual. For example, it can allow niche expansion to capture prey that would be too large, well-defended, or swift for a single individual to capture; it can increase the success rate of predation attempts (which can increase mean rate of food intake and/or reduce the variance in intervals between meals); it can allow prey captures with reduced individual input in time and energy and/or reduced individual risk of injury; and lastly (as considered in the next section) it might allow more successful defense of kills and protection from kleptoparasitism. Against this, there will be competition among group participants for the spoils of successful captures, leading to potential for reduced or even zero nutritional reward to a given individual from a successful kill (Bailey et al., 2013).

Instances of group hunting involving MSGs of hunters are very rare. The best documented case involves cooperation between predators with complementary predatory methods for capturing fish around coral reefs, studied by Redouan Bshary et al. (Bshary et al., 2006; Vail et al., 2013). The original observations were made involving interactions between the grouper and the giant moray eel. The grouper is an open-water piscivore; when prey fish detect them, they often seek shelter in crevices of coral reefs where groupers cannot access them. However, it has been observed that in response to prey fish seeking shelter, groupers search for nearby moray eels and signal to encourage them to hunt, indicating areas of the reef where prey have taken refuge (Section 6.3.1). Eels can access fish in crevices and indeed capture some of them by trapping them in dead ends. But some prey will be driven to break cover by the eels, when they again become vulnerable to the groupers. Thus, this would seem an example of interspecific cooperative hunting based on complementary hunting techniques. It is likely that both parties benefit, or else groupers would not seek out the eels and eels would not respond to signaling from the groupers. Further work has shown the coral trout show similar behavior to groupers; and the Napoleon wrasse and octopuses show similar behavior to the eels in MSGs (Vail et al., 2013). The complementarity of the hunting techniques of the two partners in these phenomena is very reminiscent of the cooperative hunting between badgers and coyotes described in Section 3.6.2.

Another instructive example of cooperative hunting is the relationship between honeyguides and humans. The research of Isack and Reyer (1989) has demonstrated that honeyguides actively recruit humans toward beehives, making distinctive calls, flying in the right direction, and indicating the proximity of the hive through a combination of signals. The birds then benefit when humans open the nests allowing them to feed on the wax honeycombs and the bee larvae. Humans can also produce signals to entice honeyguides to search for honey (Spottiswoode et al., 2016).

Bshary et al. (2006) speculate that predatory MSGs are unusual since in many potential cases one species could entirely prevent the other from accessing the spoils of all successful kills and so the subdominant species would have no incentive to cooperate. The cases discussed above may be unusual cases where such complete competitive exclusion cannot occur, or the two species consume different parts of the collective spoils, such as is the case for the honeyguides and humans.

4.2.6 Defending a Food Source

It is relatively well documented that group formation can help defend or usurp a food source (Marzluff and Heinrich, 1991; Carbone et al., 1997). Similarly, the ability to maintain exclusive access to a feeding territory is often seen as a strong driver of group formation in primates. For example, Gautier-Hion et al. (1983) presented evidence that MSGs among monkeys allowed defense of an area that lead to less need for extensive searching of the environment and an ability to more effectively exploit the richest supplies of fruit. Garber (1988) reported a similar example of species collaborating in territorial defense.

4.2.7 Improved Efficiency of Exploiting the Local Environment

It is often postulated that where resources in the local environment can be depleted by foragers and take some time to recover from such depletion, group foraging might be attractive because it reduces the likelihood of visiting a food patch that has recently been depleted by another (Cody, 1971; Terborgh, 1983). This situation was investigated theoretically by Beauchamp and Ruxton (2005), who explored the performance of such grouped foragers versus solitary individuals sharing the same environment. In general, the solitary individuals did pay a cost in recently depleted food patches that are visited rarely; however, this was usually less than the cost that the grouped individuals paid for quickly depleting any patches visited and thus spending a high fraction of their time searching for new patches to exploit. However, there may be some ecological situations where such travel costs are low: one can imagine a situation where trees that might contain fruit are very obvious in the environment and so can quickly be found, but the state of depletion (the number of fruit on the tree) cannot readily be seen at a distance and requires a visit to the tree to evaluate; thus, there may be some situations where group foraging does provide benefits through this mechanism. Furthermore, there may be particular benefits to mixed-species grouping in this context if individual species forage on different resources within the same patch (say, at different heights within the fruiting trees) and/or if they have complementary abilities to find new patches. Finally, it is worth noting that an alternative means of avoiding recently depleted patches is to be able to detect cues and signals of recent visitations; this is known to be the case in bees visiting flowers, where they are sensitive to the hydrocarbon deposits left by the feet of other recent bee visitors; such detection is known to work across species as well as within species (Goulson et al., 1998).

It might be worthwhile extending the theory discussed immediately above, to cover a situation where a number of individuals act together to defend an exclusive territory. In this context, it might be worth asking whether these individuals then perform better during foraging by remaining together or separating and all foraging separately. Finally, it might be worth considering costs and benefits in terms of continued defense of the territory; by separating during foraging, the territory holders will be spread across the territory more and so will be able to detect encroachment by others more quickly. However, the collective ability of the territory holders that are initially spread out, to repel any detected encroachers, may be reduced until such time as they reform into a coherent group. Given that many primates and bird species are often territorial but form MSGs, such theoretical investigations might usefully be teamed with empirical study (perhaps taking advantage of recent advances in on-animal data-collection telemetry, Section 9.3).

4.3 SOME POTENTIAL FORAGING COSTS OF (MIXED SPECIES) GROUPING

We will cover costs more briefly that benefits; not because they are less important, but because many of the essential concepts have been introduced when we described potential benefits.

4.3.1 Competition for Food, Food Stealing, Dominance, and Within-Group Aggression

As we described fully in Section 4.2.3, competition for food can be very complex in MSGs. However, it is clear that a subordinate species may experience a foraging cost to being in an MSG relative to a single-species group or foraging alone. In some cases, this may cause individuals to leave the group, as observed in lapwings when gulls joined an MSG with plovers in the study of Barnard and Thompson (1982). However, in other cases, MSGs may still be stable in such circumstances if: (1) the subdominant species gains a compensating nonforaging benefit, such as protection from predators, as with the plovers aggregating with the lapwings in the absence of gulls in the study of Barnard and Thompson, (2) or if it would be too costly or impossible for the subordinate species to forage away from the dominant species (as likely occurs for the marsh tits dominated by the great tits in the study of Sasvári, 1992).

4.3.2 Conflict Over When and Where to Feed and/or Over Speed of Movement Through the Environment

Conflict between males and females over optimal scheduling of feeding versus resting seems to be a strong driver of sexual segregation in large mammalian herbivores that live in groups (Ruckstuhl and Neuhaus, 2002; Ruckstuhl, 2007). Such conflict in the scheduling of behaviors has also been implicated in age-assorted grouping in sheep (Ruckstuhl, 1999) and size-assorted shoaling in fish (Aivaz and Ruckstuhl, 2011). It seems very likely that similar conflict between the different species driven by the conflicting demands of different physiologies will also act to make MSGs less attractive. In fish, although MSGs are often encountered (Krause et al., 1996), fish generally show a strong preference for grouping with size-matched individuals of the same species (Ward et al., 2002). There may be a number of reasons for this, including antipredatory benefits and reduced competition for food, but another reason may be preferred swimming speed. There is a tail-beat frequency that maximizes the energetic efficiency of swimming, and this will translate into different preferred swimming speeds for fish of different sizes and species (since the different anatomies of different species will influence the propulsive force of each tail beat and the drag experienced; Videler and Wardle, 1991). Thus, the energetics of transport may select against MSGs in fish and swimmers generally. A similar argument can be made for birds, where there is a "maximum range speed" that maximizes energetic efficiency of flight and will vary between species according to size, wing shape, and muscle physiology (Pennycuick, 1989).

Pomara et al. (2003) observed the foraging behavior of slate-throated redstarts both when solitary and when in an MSG. An increased movement rate but not an increased prey-capture rate was observed when foraging in groups. These observations were thought to reflect a switch from a sit-and-wait form of fly catching to a more active foraging strategy suitable for keeping up with the rest of a moving flock. Hutto (1988) recorded a similar phenomenon in avian MSGs in Mexican forest, with some species moving from tree to tree as part of the group at a rate quite different from that when they are alone. Hutto interpreted this as those species paying a cost of reduced foraging efficiency to buy antipredator benefits of grouping. Similarly, Darrah and Smith (2013) observed different foraging behavior in the wedge-billed woodcreeper when in an MSG compared with foraging alone, in a way that they interpreted as the woodcreeper trading-off reduced foraging efficiency for antipredatory benefits of grouping.

4.4 REDUCED COST OF LOCOMOTION IN GROUPS

There is now good empirical evidence that large birds flying in formation and fish swimming in a coordinated group of conspecifics can reduce the energetic costs of travel (Weimerskirch et al., 2001; Marras et al., 2015). At present, there is no evidence for this mechanism in MSGs. In addition, at present the mechanisms behind these energy savings are not well enough understood to allow us to make predictions as to whether there might be circumstances in which an MSG offers additional savings to any group members compared to a single-species group. Energy savings probably make more sense in situations where individuals travel over long distances. In foraging MSGs, travel is often interrupted by foraging activities and individuals are typically quite spread out, which makes energy savings unlikely. Furthermore (as we argued in Section 4.3.2), different species likely have different optimal speeds, and this will conflict with species obtaining energetic savings from MSGs.

4.5 PROTECTION FROM ADVERSE ENVIRONMENTS IN GROUPS

One selection pressure for grouping can be protection from environmental stress. Hence, emperor penguins huddle together to reduce each individual's exposure to wind and cold air; and by day, slugs often aggregate somewhere shady and form a tight ball that minimizes water loss to the warm air (Krause and Ruxton, 2002). We know of no instances where MSGs form for this purpose. However, since the benefits of such aggregation often increase with group size, species banding together may bring potential benefits of a larger group. Also, different species may have different niches and thus allow aggregation with reduced costs. For example, in the case of the slugs, those that inhabit a mixed-species ball by day may subsequently experience less competition from ball mates after the members of the ball disperse as the sun goes down.

In contrast to animals, the physical effects of mixed-species assemblages are well studied in plants (Callaway, 1995). The altering of the physical environment by neighboring individuals is a much more pressing concern for plants than animals, given their very limited scope influencing their positioning with respect to others. Callaway documents situations where individuals of one plant species are often found more closely associated with those of another species than predicted by chance, and experimental manipulations allow ruling out explanations based on shared microhabitat preference and mechanisms such as one species having a physical structure that naturally captures the seeds of another. Benefits of such an association in terms of exposure to environmental stress involve protection from temperature extremes and modification of soil physics and chemistry. For example, especially in desert environments, some plants often only thrive in the shade of another, and this shading protects against very high surface temperatures that could cause damage directly or cause excessive water loss. Other benefits include protection against excessive sunlight leading to photoinhibition, wind causing physical damage, and abrasion from suspended particles. In this situation, there will also be costs to the shaded species in terms of reduced photosynthetic ability when sunlight is weaker and limiting. There are also likely to be costs to both parties to close proximity, most obviously in terms of competition for water and nutrients through the root system. There can also be soil-based benefits with the root system of one species breaking up compacted ground in a way that benefits another. If surface soils are nutrient poor, then a deep-rooted species may be able to flourish, and leaves shed by that species may add sufficient nutrients to top soils to facilitate growth in a shallower-rooted species.

Generally, the costs and benefits to both parties of such associations are complex and not fully resolved for any system. The fact that plants are sedentary for all but the seed stage and seeds are not self-directing reduces potential for evolutionary selection of traits to facilitate association. However, seed germination is influenced by a range of cues, and it is not difficult to imagine how the shade-seeking species could evolve to germinate preferentially when detecting cues suggestive of a shaded microhabitat. If the shade-providing plant suffers strongly from competition from shade seekers, then again it seems possible that it could evolve to biochemically influence the surrounding soil in a way that inhibits germination of potential competitors—but such adaptations are (to our knowledge) unexplored.

4.6 FURTHER SOCIAL AND REPRODUCTIVE ASPECTS OF (MIXED-SPECIES) GROUPING

These have not been extensively studied: the brief review on mammals MSGs offered by Stensland et al. (2003) remains the key work in this area. Stensland et al. suggest that in territorial species (most obviously primates) the ability to defend a large home range against neighboring groups is strongly determined by numbers, and formation of MSGs may be an effective means of achieving this for members of several species simultaneously. This may be particularly effective when the species in the MSG have separate foraging niches, and so an MSG may be even more attractive to all parties than a same-sized group of conspecifics. Also relevant to group formation for territorial defense, but also more widely (see Section 4.2.4), individuals in an MSG may experience less aggression related to breeding opportunity, so even if there is competition for food between species in an MSG, there may still be foraging advantages of forming an MSG, if this allows more time to be devoted to feeding because less time is spent in aggression, vigilance for potential aggression, and appeasement behavior toward high-status individuals. An alpha male in a group with females that he wants to maintain sexual exclusivity with may be more comfortable associating with heterospecific males (Buchanan-Smith, 1999). This situation is complicated by the potential for hybridization, which occurs in at least 10% of animal species (Willis, 2013) and is suspected in species that often form MSGs (Li et al., 2010; Hodgins et al., 2014).

A very unusual case of reproductive benefits within an MSG was reported by Schlupp et al. (1994). The Amazon molly features female gynogens that reproduce clonally but rely on sperm from a heterospecific to initiate embryogenesis. It was originally assumed that male sailfin mollies mated with these females by mistake and gained no benefit from the interaction. However, Schlupp et al. demonstrated that female sailfin mollies show a preference for mating with conspecific males if they have observed those males successfully mating with Amazon mollie females. The underlying selective advantage for this mate choice copying across species has yet to be identified.

Finally, Stensland et al. (2003) speculate that young or subordinate individuals may benefit from being able to practice mating behaviors on another species within MSGs with reduced likelihood of experiencing aggression from dominant males because (in the absence of risk of hybridization) there is little cost to the dominant males of allowing such behaviors from heterospecifics—making investment in driving them off unattractive. However, this idea has yet to be extensively explored.

There is evidence that migrant passerine birds preferentially nest beside resident species as a means of selecting good breeding territories. This has been most extensively studied in the pied flycatcher, which (conveniently for researchers) uses nest boxes (Seppanen et al., 2011). As discussed in Section 2.3.1, birds of this species preferentially use nest boxes in close proximity to those in use by resident great tit and blue tits, and show a stronger preference for association with individuals that show (e.g., by large clutch size) that the local environment is food rich. This behavior has been demonstrated to both benefit the flycatchers and incur a cost to the tit species. However, since the tits typically lay their eggs before the flycatchers begin nesting and are not sufficiently aggressive to drive flycatchers away, there is little they can do to combat this information parasitism.

Mixed-species grouping can be beneficial for pollination and seed set on animal-pollinated plants but only in rather specific circumstances (Feldman et al., 2004; Peter and Johnson, 2008). One species might be more attractive to pollinators than another (because it has more conspicuous and/or rewarding flowers), and this has the potential to improve the pollination of the less attractive species when in an MSA, but this benefit is not guaranteed. If the pollinators show very high levels of floral constancy, then those drawn to the association by the more attractive species will not visit the less attractive species and so there will be no benefit. If, on the other hand, there is very low floral constancy, then being in an MSA may be unattractive because of the higher likelihoods of alien pollen being brought to a focal flower or pollen taken from the focal flower being deposited on the other species. However, it is possible to imagine an intermediate between these extremes where a benefit is possible. Costs and benefits will vary with the relative abundance of the species concerned and with the focal pollinator community, and so there is unlikely to be a sufficiently consistent selective benefit to drive strong evolutionary adaptation in this regard.

4.7 CONCLUSIONS

We began this chapter by delineating a range of different situations that can occur with respect to the net fitness consequences of mixed-species grouping. We then emphasize a further level of complexity, where the net fitness consequences for one party in an MSG can involve a number of (potentially interacting) underlying mechanistic drivers; and those mechanisms might be quite different for members of different species within an MSG. We then go on to scrutinize these different potential mechanisms, dealing first with foraging.

An overarching theme in mixed-species grouping is tension involving extent of niche overlap between species. Specifically, some of the foraging benefits of an MSG (such as local enhancement) require some overlap in niche; however, the greater the overlap, the greater competition between different members of an MSG will be. Hence, we expect that there will be no simple relationship between niche overlap and the species composition of MSGs.

We argue that we expect local enhancement (at a range of spatial scales) and beater effects (or enhancement of food availability or accessibility more generally) to be widespread benefits to MSGs. It may initially seem that reduced feeding competition is the most obvious foraging benefit associated with mixed-species grouping; however, we argue that the costs and benefits associated with competition can be subtle and complex in mixed-species grouping. An important aspect of this is that competition between members of one species will often be influenced by close proximity of members of another species. We highlight reduced variance in intervals between meals and cooperative defense of an exclusive foraging territory as potential benefits of mixed-species grouping that might particularly merit further research attention.

We argue that physical benefits (in terms of energetic costs of locomotion or protection from adverse environments) do not seem a strong driver of mixed-species grouping, even though they are important in diverse single-species associations. We briefly highlight plants as an interesting counterpoint, where physical benefits of an MSA have been well documented in diverse systems. Similarly, reproductive mechanisms do not seem strong underlying drivers of MSGs, except in unusual circumstances.

Chapter 5

Adaptive Implications of Mixed-Species Grouping: Predators and Other Antagonists

This chapter is the complement to the previous chapter by focusing on costs and benefits of mixed-species groups (MSGs), whereas the main focus of the previous chapter was on foraging. Here we are chiefly interested in predator avoidance. Having explored a range of different mechanisms by which MSGs in animals can affect their risk of predation, we expand our consideration for protection of animals from parasitism and protection of plants from herbivory. We finish the chapter by focusing on a number of well-examined case studies of mixed-species associations (MSAs)/MSGs to see how the different mechanisms introduced in this chapter and the previous chapter interact in diverse systems. As in the previous chapter, we draw examples very widely to explore a broad range of relevant mechanisms.

5.1 CONSEQUENCES OF MIXED-SPECIES GROUPING FOR PREDATION RISK

5.1.1 Encounter-Dilution Effects

If predators need to search their environment for prey, then associations will influence the rate at which prey are discovered and targeted by predators. In particular, if predators can only capture one individual when they discover a group and the rate at which a group is discovered by predators increases less than proportionally with group size, then per capita predation risk for the prey is reduced by association; this is the *encounter-dilution* or *attack abatement* effect, as coined by Turner and Pitcher (1986). This should be a widely acting antipredatory benefit of grouping. The benefit can continue to occur even if more than one individual from a group is captured in one attack, as long as not all group members are captured. The available evidence is that a group of N prey is generally not N times more easily detected (and thus N times as frequently attacked) as a single individual, although empirical research on this is relatively modest (Ioannou and Krause, 2008). Furthermore, even if these assumptions do not hold and association has no effect on the mean rate at which a predator captures prey, it should increase variation in the predator's time between prey encounters. This should reduce the attractiveness of the prey to predators, who risk starvation (or other costs of being in poor condition) during long periods without food, perhaps causing them to switch to other prey or to other areas.

There is potential for an added encounter-dilution benefit of mixed-species grouping if MSGs are inherently less detectable that single-species groups of the same size and conversely a potential cost if MSGs are more detectable. However, this remains entirely unexplored, both theoretically and empirically. Some species may be inherently less detectable at a distance (because they are cryptically colored or forage quietly or produce low volatile emissions), and grouping with this species may offer enhanced encounter-dilution benefits. But even if both species are equally inherently detectable and if they are differentiable at a distance by predators, then it may be that an MSG is less detectable than a similarly sized single-species group. If we consider visual detection, then it seems possible that a heterogeneous mixture of two species has less visual salience than a similar-sized homogenous group, because of the greater complexity of appearance of individual groups and greater variation in appearance between groups.

If we turn to the issue of what fraction of individuals in a group are consumed when that group is discovered by a predator, then again there is potential for a benefit to accrue to one species from association with another. Say if one species is less palatable or is even aversive, then we can imagine that a predator might break off its consumption of individuals in a group more readily after encounters with this more aversive type, so a less aversive species might experience enhanced encounter-dilution benefits from associating in an MSG with a more aversive species. In support of this, De Wert et al. (2012) demonstrated that predation rates on palatable pastry baits by free-living wild birds were reduced if these baits were placed in groups with visually different aversive tasting baits. Given encounter-dilution's likely commonness as a factor in the evolutionary ecology of aggregation and the reasons to expect it to be modified by species composition in an MSG setting, further exploration of encounter-dilution involving more than one species is very much warranted.

5.1.2 Vigilance and Collective Detection

There are obvious benefits to grouped prey if an imminent predatory threat detected by one individual in the group is communicated to all. Such collective detection can either allow reduced individual predation risk or allow reduced investment in antipredatory vigilance without increased predation risk. The evolutionary ecology of this has been intensively studied for decades (Beauchamp, 2014). These benefits can apply to MSGs, provided the species concerned share at least some predators in common and provided that responses of an individual of one species to a predator can reliably be detected by individuals of the other species. The first of these conditions is commonplace but not ubiquitous.

There may be little overlap in predators if the species involved in an MSG differ dramatically in size and/or are not closely related. There may also be reduced overlap of predatory risk if there is spatial segregation within the group (with a tree-dwelling species being exposed to aerial predators and ground-dwelling associates being exposed to terrestrial predators). The second condition is also commonplace, with examples of predator-specific alarm calls being interpreted similarly by conspecifics and heterospecifics (Griffin et al., 2005). Clearly from this discussion, it can be seen that there may be circumstances where associating with heterospecifics should bring vigilance benefits. However, these benefits will often not be as strong as would be gained by associating with a similar number of conspecifics because of differences between species in what predators are threats in the speed with which a detected threat changes the behavior of the individual and/or in the ease with which such behaviors can be detected.

However, there are also circumstances when an MSG can offer benefits over a similar-sized single-species group. Such a situation occurs where the species have complementary predator-detection abilities that can be combined to offer more effective collective detection. This may come about because of spatial segregation in the group. This has been extensively explored in MSGs involving species of Amazonian primates (Heymann and Buchanan-Smith, 2000). Often in such primate groups one species will feed characteristically higher in the trees than the other; both species are at risk from aerial, arboreal, and terrestrial predators that approach on the ground but can scale trees. The higher feeding species devotes more attention to arboreal and aerial threats and is more effective at detecting them, communicating such detections to individuals of both species. The same is true with respect to the lower feeding species and ground-approaching predators. Individual investment in vigilance can decrease in MSGs and vigilance can become more spatially focused in MSGs, for example with the lower feeding group devoting less effort to upward-directed scans.

Arboreal feeding associations between red colobus, olive colobus, and Diana monkeys have been extensively studied (Stensland et al., 2003). It would seem that the colobus monkeys initiate formation of the MSG and benefit from Dianas' tendency to feed on the outermost branches of trees offering them clearer lines of sight for detecting approaching (general avian) predators. Dianas may still benefit through enhanced vigilance in a larger group because the colobus monkeys still have some ability to detect approaching threats.

Benefits may derive from different intrinsic sensory or other physiological abilities between species in resting MSGs involving cowtail stingrays and whiprays (Semeniuk and Dill, 2006). Whiprays responded earlier to simulated predator attacks, and this may be related to mechanosensors in their tail. In support of this, cowtails preferentially settle with whiprays with longer tails. In the partnerships between fishes and burrowing shrimps, individuals of both species share a burrow that the shrimp constructs, but both must leave the burrow to feed. The fish has much superior vision and alerts the shrimp to potential predatory threats allowing both to retreat to the burrow (Karplus and Thompson, 2011). Norris and Dohl (1980a) speculated that spinner dolphins seek out spotted dolphins to rest in proximity to a species that shares predators (e.g., sharks and killer whales) but is intrinsically more alert. The evidence in support of this theory remains circumstantial: the two species do not forage together, and in the absence of spotted dolphins, spinners rest in open clear-water habitats where approaching predators should be easier to detect. Goodale and Kotagama (2005a) studied avian MSGs in Sri Lanka, where individuals responded to heterospecific alarm calls. They found that one species generally alarm called earlier to real predatory threats but was also more likely than the other species to alarm call in response to nonpredatory stimuli. They speculated that both species might benefit from integration of different characteristics of the two species' alarm systems. For example, it might be effective to raise vigilance levels in response to calling of the early responding species but only initiate more active fleeing responses if there is subsequent calling by the more reliable species.

Recently Schmitt et al. (2016) demonstrated that zebras reduce their vigilance more when herding with giraffes compared to conspecifics and direct their vigilance more toward giraffes than to the wider environment (Fig. 5.1). The authors speculate that the height of giraffes makes them very adept at detecting approaching threats (with lines of sight being less obstructed by vegetation) and also makes responses (freezing and staring) triggered by approaching threats very readily seen by nearby zebras.

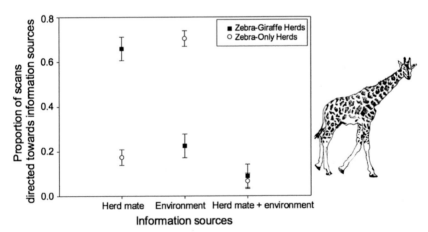

FIGURE 5.1 The apparent targets of zebra visual scans. The mean (±1 standard error) proportion use by zebras of both herdmates or the wider environment as an information source about approaching threats, as determined by the direction of their gaze. *Adapted from Schmitt, M.H., Stears, K., Wilmers, C.C., Shrader, A.M., 2014. Determining the relative importance of dilution and detection for zebra foraging in mixed-species herds. Animal Behaviour 96, 151–158, and provided by Melissa Schmitt; sketch by Isabella Hoskins.*

5.1.3 Confusion and Oddity

The confusion effect is the observation that attacks by predators are less likely to be successful when the attack requires the predator to single out and track one specific individual from a group of moving prey. This effect is considered to be a consequence of the predator's inability to accurately process all the visual stimulation generated by a number of moving potential targets; its theoretical and empirical aspects are reviewed by both Krause and Ruxton (2002) and Beauchamp (2014). The oddity effect is the observation that individuals whose appearance is different from the majority of group members are more likely to be targeted by predators. This is generally seen as a means for the predator to reduce the confusion effect, but it may also be that minority individuals in a group are just more salient (i.e., that unusual prey simply draw the predator's attention more). Because species in an MSG often differ dramatically in group size, body size, appearance, and behavior, the oddity effect should be particularly relevant in such groups, imposing a potential cost to species that make up the minority of a group.

These effects of mixed-species grouping on predators' targeting were explored theoretically in a multispecies setting by Tosh et al. (2007), who used an artificial neural network to model the retinotopic mapping that seems to be a key component of visual attention across animals that hunt by sight. They predicted that visual targeting of a species with a cryptic appearance was hindered when that species associated with another species of more conspicuous appearance. However, the benefit was found to be asymmetric because targeting of conspicuous individuals was more hampered in uniform groups than in MSGs. This prediction was tested by Rodgers et al. (2013), using two different types of *Daphnia* water fleas predated by stickleback fish. They found that cryptic individuals were targeted less often than predicted by chance when they were grouped with a lower or equal number of conspicuous individuals. In this system, a single conspicuous individual in a group of conspicuous individuals was targeted randomly. A cryptic individual's risk of being targeted for attack was greatest in a uniform group and declined as the fraction of the group that was conspicuous increased; in contrast for conspicuous individuals, risk was lowest in a uniform group and increased as the number of the group that was cryptic increased (Fig. 5.2). These results provide support for the theory of Tosh et al., but to explore the likely consequences for mixed-species grouping, we would need to explore variation in group size and composition (when an individual joins a group, it changes that group's size and composition). It would also be useful to explore how variation in group size and composition affected encounter-dilution and prey targeting within the group. However, the theory of Tosh et al. and the experiments of Rodgers et al. do suggest that the oddity effect may not hold universally, but that nonrandom targeting within MSGs may well be important.

There is further empirical evidence related to oddity and MSGs. Almany and Webster (2004) compared patterns of species richness of shoaling juvenile

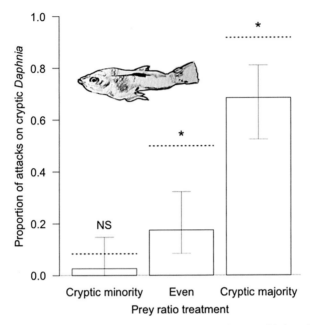

FIGURE 5.2 Exploring the oddity effect in laboratory experiments with foraging fish. The proportion of attacks on cryptic *Daphnia* by three-spine sticklebacks in the three different prey ratio treatments (ratio of cryptic to conspicuous individuals for each treatment is as follows: 1:11 Cryptic minority, 6:6 Even, 11:1 Cryptic majority). *Dotted lines* indicate the expected proportion of attacks targeting a cryptic *Daphnia*, based on random expectation. *Adapted from Rodgers, G.M., Kimbell, H., Morrell, L.J., 2013. Mixed-phenotype grouping: the interaction between oddity and crypsis. Oecologia 172, 59–68, and provided by Lesley Morrell; sketch by Isabella Hoskins.*

fish on reefs from which predators had either been experimentally removed or not and concluded that the most likely explanation for the patterns observed was predators preferentially capturing rare species. In follow-up laboratory experiments, preferential targeting of rare species in groups was demonstrated (Almany et al., 2007).

The most extensive empirical study of the oddity effect was performed by Landeau and Terborgh (1986) and used normal silver and blue-dyed prey fish (minnows) attacked in the laboratory by a large predator fish (bass). There was an evidence for an oddity effect in that in a group of eight if one or two individuals were of a different morph from the rest, then the likelihood of an attack on the group being successful was considerably higher than if numbers were equal or near equal, and the attacked individual was disproportionally likely to be of the rare morph. However, a lone individual was almost always captured, whereas a lone individual who joined a group of seven heterospecifics would only be killed in around 50% of trials. Moreover, an individual in an MSG of four silver and four blue minnows was less vulnerable when in a homogenous group of four and only slightly more vulnerable in a monomorphic group of eight; so the oddity effect shown in this experiment might not be enough to select against MSGs. Furthermore, having one or two odd individuals in a group of 15 did not increase a predator's ability to make captures compared to a monomorphic group of 15, suggesting that the oddity effect was swamped by the confusion effect. The authors conclude that a lone individual would benefit from joining an MSG and that individuals of the majority phenotype would find such joining a positive or neutral experience, except when the group size is small and group composition strongly imbalanced.

There have been a number of other demonstrations of oddity effects in field and laboratory experiments. Rutz (2012) showed that birds of prey preferentially targeted rare color morphs in feral pigeons, which are generally found in groups. Allen and Anderson (1984) offered two colors of pastry baits to wild-living birds and showed that at low densities the common morph was preferentially taken (perhaps because of search images effects), but at high prey density the rare morph was preferred (perhaps due to oddity effects even when the prey were immobile). Wilson et al. (1990) offered two colors of live maggots on a bird table, the temperature of which could be controlled: at higher temperatures the maggots moved more and free-living birds' preferential targeting of the rare morph increased. But the most relevant evidence in a multispecies context is indirect: Wolf (1985) found that rare fish often left MSGs when cues of predator were present and were increasingly likely to do so, the rarer they were in the MSG. However, the empirical evidence is not entirely unequivocal: Caldwell (1986) found no evidence that darker birds were targeted more often in groups mostly consisting of lighter colored individuals.

5.1.4 Mobbing and Aggressive Defense

One species can gain protection from their predators by association with a more aggressive species that can either deter predators from approaching or drive them away (Section 2.4.1). This has been most extensively studied in birds nesting in association with another more aggressive bird species or with social wasps (Quinn and Ueta, 2008). Although there is no reason why these benefits need to be restricted to nesting (Beauchamp, 2014), they (in terms of fledging success) are easily quantified in this context.

Quinn and Ueta demonstrate that certain avian species nest more commonly with more aggressive birds or wasps than expected, and that such association brings reduced predation risk. Numerous studies have shown that the strength of these benefits can vary with proximity to a nest of the aggressive species and with individual protector aggressiveness (Quinn and Ueta, 2008). For example, Bogliani et al. (1999) studied the benefit that wood pigeons gain from nesting near the Eurasian hobby. They showed that wood pigeons preferentially nested near more aggressive hobby pairs, and that experimental dummy nests experienced reduced egg predation the more aggressive the individuals in the nearest hobby nest.

There can be costs as well as benefits to the "protected species" with the most commonly reported costs being offspring predation by the more aggressive species. In the study cited above, Bogliani et al. reported that 15% of prey weight at hobby nests consisted of woodpigeon chicks. Presumably this cost to wood pigeons is outweighed by reduced predation due to other predators that the hobbies effectively deter. Nesting in association with the more aggressive species may also mean nesting in habitats that are suboptimal in other ways. Bijlsma (1984) demonstrated that woodpigeon nests were more commonly destroyed by high winds when associating with hobbies, whereas nonassociating wood pigeons selected habitats that offered shelter from the wind. Prop and Quinn (2003) demonstrated greater competition for food when nesting near to a preferred aggressive associate in red-breasted geese.

Quinn and Ueta conclude that many such associations are commensal with the more aggressive species being unaffected. However, they discuss one study where the aggressive species (the black skimmer, large billed terns, and yellow-billed terns) suffered reduced reproductive success from association with sand-colored nighthawks (which benefitted from the association). Skimmers and terns spent more time defending the colony from predators when associated with nighthawks (presumably because the presence of the nighthawk nests attract more predators, which were predominantly birds of prey), and the reduced parental attentiveness lead to reduced reproductive success (Groom, 1992). It may be too costly for the other species to attempt to drive away the nighthawks from mixed-species colonies, because nighthawks often outnumber the other species 10–1. It may also be too costly for the more aggressive species to move to an alternative breeding site to escape the nighthawks, because nighthawks typically lay in a quick-to-make nest scrape a few days after the more aggressive species. Thus the more aggressive species could only move away from nighthawks at the cost of abandoning their clutch. Hence, this may be a case where the aggressive species has no better option than to accept the cost of nesting with the protected species.

Another system where there is evidence of an impact on the presumptive protector species involves lesser kestrels nesting in association with less aggressive jackdaws (Campobello et al., 2011). Here both species seem to obtain a benefit in terms of reduced investment in antipredatory vigilance compared to same-sized single-species groups. Furthermore, lesser kestrels showed reduced strength of mobbing response to presentation of model predators, while jackdaws in small colonies increased their rate of alarm calling in mixed-species colonies. The authors interpret this as a mutualism that allows each party to gain valuable information from the other about predatory threats: lesser kestrels spend more time in the air and this may provide them greater ability than jackdaws to detect predators from a distance, whereas jackdaws have a complex alarm call repertoire that may allow them to encode very specific information about the nature of predatory threats. The authors argue that responsiveness to heterospecific alarms allows both species to maintain effective antipredatory responses at reduced individual investment. However, this explanation remains speculative, and the consequences of this MSA in terms of measures of reproductive output remain unexplored. However, this is another promising system for exploration of the complexity of mixed-species nesting associations.

These benefits of proximity to an aggressive species are not restricted to nesting birds; see Section 2.4.1 for the diversity of taxa engaged in protective associations. For example, Le Guen et al. (2015) report on a nesting mutualism between an arboreal ant and a social wasp species. The presence of the wasps reduces avian predation on the ant, whereas the ant was able to divert army any raids on the shared host tree that would otherwise pillage the wasp nest.

5.1.5 Nonrandom Prey Choice Within Groups by Predators

This mechanism is only relevant if predators can only capture a subset of the prey group (often only one individual) any time a group is attacked. If prey species share the same predator species, but the predator has a preference between the two, then the less-preferred species might gain a benefit from grouping with the more-preferred species. This should reduce the fraction of times that an individual of the less-preferred species is targeted during attacks by the predator. However, it is by no means guaranteed that this mechanism will bring a benefit to individuals of the less-preferred species. This benefit might be overridden if predators are more likely to attack groups that contain individuals of the more-preferred species; then the overall predation risk to individuals of the less-preferred species could increase if the effect of greater rate of attack on MSGs outweighs the effect of reduced likelihood of being targeted in any one attack. This trade-off has not been explored either theoretically or empirically to our knowledge. There may also be other nonpredatory reasons, such as food competition, why this mechanism is not sufficient to make MSGs attractive to the less-preferred species. For example, Pays et al. (2014) found that impala in monospecific groups showed a decrease in vigilance with increasing group size, but no change in vigilance with group size in MSGs, an effect that they put down to intensity of food competition in MSGs (Fig. 5.3).

Of course, this mechanism also may make MSGs risky situations for the more-preferred species, although it need not necessarily dominate other selection pressures related to mixed-species grouping. For example, predation risk could sometimes decrease for a member of the more-preferred species if it joins a group of less-preferred species to make an MSG. This can occur if the effect of preferential targeting is overwhelmed by predators attacking MSGs less often or if the likelihood of an attack being successful is reduced by the presence of members of the less-preferred group. Again, we know of no theoretical and little empirical work exploring the antipredatory consequences of selectivity for members of the more-preferred species. Such a study would ideally investigate the consequence of being in an MSG versus being alone and being in a single-species group.

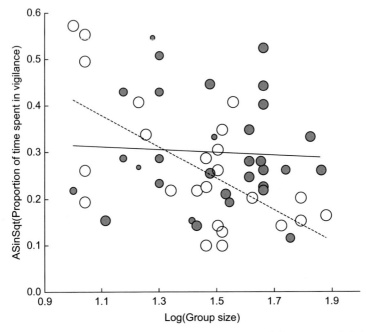

FIGURE 5.3 The effects of group size and composition on vigilance rates. Relationship between the arcsine square root-transformed proportion of time an impala spent in vigilance and the log-transformed group size in impala groups (*gray dots*) and mixed-species groups (MSGs) (*open symbols*). Dot size varies with the proportion of impalas in MSGs. In monospecific groups but not MSGs, vigilance declined with group size. *Adapted from Pays, O., Ekori, A., Fritz, H., 2014. On the advantages of mixed-species groups: Impalas adjust their vigilance when associated with larger prey herbivores. Ethology 120, 1–10, and provided by Olivier Pays.*

The most instructive study on this effect is on two freshwater sympatric fish species that share predators: armored brook sticklebacks and unprotected fathead minnows. When Mathis and Chivers (2003) exposed MSGs to predatory fish, minnows were preferentially targeted. The shoaling preferences of both prey species were studied, both in control conditions and when perception of predation risk was enhanced by addition of alarm compounds in the water. Minnows preferred to group with conspecifics over heterospecifics, even when predation risk was high. In contrast, sticklebacks preferred to associate with conspecifics when predation risk was low but heterospecifics when it was high. In a foraging study, minnows were shown to be more efficient foragers than sticklebacks, and the authors suggest that this explains the preference for conspecific grouping by sticklebacks when predation risk was perceived to be low, to reduce competition for food. However, they suggest that grouping with preferentially targeted minnows may override oddity and food competition factors when perceived predation risk is high. A surprising aspect of this study, however, was that although predators preferentially targeted minnows, they were actually more successful when attacking sticklebacks; the authors suggest that this is an artefact of their experimental arena offering nowhere to hide and thus allowing repeated attacks. In nature, they suggest that most initial attacks would occur on minnows, allowing other group members to flee to a place of safety. Over the 16 h of the experiments repeated attacks may have led to prey fish being unnaturally exhausted, and this may have artificially increased the vulnerability of sticklebacks.

Another relevant study is that of FitzGibbon (1990) on naturally occurring cheetah attacks on Thomson's and Grant's gazelles that often form MSGs. When cheetah attacked MSGs, they nonrandomly targeted the smaller Thomson's gazelle. We return to this study system in Section 5.4.1. Schmitt et al. (2014) also report on MSGs of African ungulates. They found that plains zebras reduced their vigilance substantially when in MSGs with blue wildebeests. However, strangely, the drop in vigilance was insensitive to the relative or absolute number of wildebeests in the group. To us, the most plausible explanation of this is that lions preferentially target wildebeests over zebras, so association with any wilde-beest significantly reduces predation risk to zebras. Schmitt et al. also reported that zebras reduced their vigilance when in an MSG with impalas but only when impalas comprised more than 75% of the group. This might be because impalas offer some but relatively small dilution and shared vigilance benefits because the main predators of zebras and impalas are different. Lion represents the greatest threat to zebras but rarely attempt to catch the fleeter, smaller impalas; impalas are however targeted more than zebras by hyenas and hunting dogs. However, our interpretation of these vigilance effects remains highly speculative until data on prey selection by predators upon such groups can be collected.

5.1.6 Educating Naive Individuals About Predatory Threats

It may be beneficial to learn about predators rather than have a genetically encoded innate knowledge in circumstances where the predator community faced by a prey individual is variable and unpredictable. The ability to learn about the local predator community provides the ability to respond effectively to new predatory threats (say from invasive species). It should also reduce false alarms and other costly behaviors caused by mistaking benign individuals for predator types that are not currently a threat. However, there are costs to learn-ing about predators rather than having innate knowledge. Firstly, individuals will likely be at increased risk of predation before they have acquired sufficient knowledge, and the experiences required for learning (often involving close proximity to a predator) are themselves dangerous.

While this attraction to social learning about predators has been documented in a range of taxa between members of the same species, it has also sometimes been documented in a mixed-species grouping context. For example, this was demonstrated in the laboratory by Mathis et al. (1996). Predator-naive brook sticklebacks gave a fright response to chemical stimuli from northern pike when paired with predator-experienced fathead minnows but not when alone or paired with predator-naive minnows. Those sticklebacks then performed that same fright response to the same stimulus in subsequent trials when alone, demonstrating that learning had occurred and been retained. These three species coexist in the wild. Ferrari and Chivers (2008) took an essentially similar approach to Mathis et al. (1996), and this time demonstrated learning of fright response to odor cues of a predatory salamander socially acquired between individual tadpoles of two different frog species (boreal chorus frog and wood frog). Manassa et al. (2013) performed similar laboratory experiments to demonstrate predator learning between two species captured as part of the summer recruitment pulse of larval fish to coral reef communities of Australia. Because coral reefs are highly variable in their predator communities, there should be advantage to these fish in learning about predatory threats quickly after recruitment to the reef. However, the communities of potential prey fish are also complex within reefs and diverse between reefs, so simple coevolved social learning adaptations between particular pairs of species are unlikely to evolve. These authors were able to show socially acquired predator recognition between species even in such complex ecosystems. Earlier, Vieth et al. (1980) had demonstrated that the European blackbird could be experimentally induced to learn to recognize a stimulus as dangerous, if exposed to that stimulus paired with heterospecific mobbing calls from species that it commonly shares habitat with chaffinch, great tit, and European nuthatch. More recently, Magrath et al. (2015b) demonstrated that wild superb fairywrens can learn to recognize and behave appropriately to previously unfamiliar alarm calls.

Being able to learn about potential predators from heterospecifics as well as conspecifics might simply be selected because this offers an increased rate of encounter with already-educated potential tutors and thus allows more opportunities to learn and so quicker learning. However, there are specific costs and benefits to learning from a heterospecific rather than a conspecific. First of all, different species may share some predators but also have some nonshared predators, and so information learned from heterospecifics might in general be expected to be less reliable than that learned from conspecifics. However, a complication in this is that one can also imagine circumstances where a given heterospecific is more reliable as a tutor than a given conspecific. Imagine two fish species, each of which experiences a range of gape-limited predators: a heterospecific of the same size may give more useful information about relevant predator threats than a conspecific that is quite different in size to the focal individual. Individuals of some species may also simply be more likely to be knowledgeable about predators than individuals of another species. For example, we can again imagine two fish species that differ in how widely they explore the environment. Individuals of the more timid species may benefit particularly from learning from individuals of the other species, if the wider roaming of these individuals makes them more knowledgeable about the range of current predatory threats. In some circumstances, heterospecifics might not be viewed as competitors to the same extent as conspecifics and thus provide better opportunities to learn if that can more readily be tolerated in close proximity. None of these issues has yet been explored empirically nor has there been theoretical exploration of how these different issues might trade-off against each other and affect selection for heterospecific social learning about predators.

5.2 OTHER ENEMIES (PARASITES AND DISEASE)

There can be antiparasitic benefits to nesting beside a more aggressive species. Clark and Robertson (1979) described a situation where individuals of a bird species suffered reduced brood parasitism by nesting in close proximity to a more aggressive species that shares the same avian brood parasite. Similarly, Smith (1968) demonstrated that several passerine species suffered less from parasitism from botfly larvae if they nested near hymenopteran nests.

In general for animals that have parasites with highly mobile stages, there should be benefits to association in terms of encounter-dilution, and all the issues raised in Section 5.1.1 with respect to predators should also hold for host-seeking parasites (Patterson and Ruckstuhl, 2013). Also, susceptibility to disease and parasitism may be influenced by group size. For example, individuals in smaller groups may be food stressed or individuals in larger groups may experience more aggression leading to social stress, and these stresses may enhance susceptibility to parasitism and disease (Leclaire and Faulkner, 2014). If being in an MSG affects these sources of stress, then this may have an impact on risk from parasites and disease.

In general, individuals in larger groups should suffer more from pathogens that spread by contagion (Patterson and Ruckstuhl, 2013). There may be benefits from mixed-species grouping in terms of the control of parasites and diseases that are spread by contagion, providing there is differential susceptibility between the species. For example, consider species A that suffers from a contact-spread parasite but which can form MSGs with species B that is not a viable host for the parasite. For an uninfected individual of species A, there should be reduced risk of parasitism from being in an MSG of size N than in a single-species group of the same size, because within-group encounters with conspecifics should decrease in the MSG and because effective spread of the parasites is hampered by unsusceptible individuals.

Evidence for these ideas about MSGs providing antiparasite protection comes from a field study of two closely related riverine guppy fish. Both suffer from ectoparasite species that are spread by close contact; however, the parasites are all host species–specific and those that thrive on one of the study fish species do not thrive on the other. Dargent et al. (2013) found that fish of both species in naturally occurring MSGs had lower parasite loads than those occurring in same-sized single-species groups (Fig. 5.4). Confounding factors cannot entirely be ruled out at this stage, and laboratory manipulation of group composition would be a valuable follow-up to this work. It would also be interesting to explore how

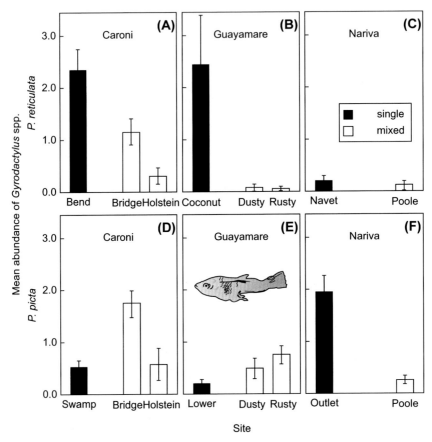

FIGURE 5.4 The influence of mixed-species association on parasite loads. Mean abundance (±1 standard error) of *Gyrodactylus* parasite spp. on *P. reticulata* (A–C) and *P. picta* (D–F) fish in single- and mixed-species sites from three rivers in Trinidad. *Adapted from Dargent, F., Torres-Dowdall, J., Scott, M.E., Ramnarine, I., Fussmann, G.F., 2013. Can mixed-species groups reduce individual parasite load? A field test with two closely related Poeciliid fishes (*Poecilia reticulata *and* Poecilia picta*). PLoS One 8, e56789, and provided by Felipe Dargent; sketch by Isabella Hoskins.*

both an individual's own infection status and the prevalence of the parasite in the wider population influence individual decisions to join groups of different sizes and composition and how parasite prevalence impacts the nature of mixed-species grouping at a population level. One of these guppy species at least is known to be sensitive to infection status when deciding on joining conspecific groups. Recently, Gonzalez et al. (2014) documented the presence of hemoparasites in more than 2000 birds from 246 species caught across Colombia. They concluded that individuals from species known to form MSGs were more likely to be infected. It is not clear, however, whether this suggests that MSGs promote infection for any other reason that being larger than single-species groups, because characteristic group size was not considered as a variable in their study.

5.3 HERBIVORY IN MSAS OF PLANTS

Many of the mechanisms discussed thus far for animals in MSGs have also been studied in relation to how plants in close association to individuals of other species are attacked by herbivores. There have been several recent reviews of this literature (Barbosa et al., 2009; Rautio et al., 2012; Underwood et al., 2014; Ruttan and Lortie, 2015). This section draws heavily on these reviews. While there are similarities in mechanisms to animal groups, there are also important differences because plants cannot move and behaviorally control which other types of plants are in their immediate neighborhood. This likely reduces selection pressure on specific traits that exploit benefits stemming from MSAs because the offspring of a plant in a given type of MSA could be no more likely to end up in an MSA themselves than those of another plant of the same species that existed in a quite different neighborhood. Nonetheless, we do see strong parallels with the mechanisms considered previously in this chapter.

Some unpalatable plants may camouflage the existence of neighbors, either physically or by means of volatile emissions. Other unpalatable plants may be so aversive as to cause herbivores that sample them to stop feeding for a while and/or leave the vicinity, again benefiting more palatable neighbors. Herbivore attack on one plant may cause release of volatiles that induce neighbors (even of different species) to produce increased chemical defenses. One plant may attract natural enemies (e.g., parasitoids) of the herbivores to the general vicinity by means of volatile signals of herbivore attack or by offering food such as extrafloral nectar; and this high local density of enemies of herbivores may deter herbivores from the vicinity or see their numbers reduced by direct attack by their enemies. Alternatively, the attraction and retention of natural enemies may be more localized to a specific plant, thus causing herbivores to flee to nearby plants, triggering associational susceptibility.

The concept of associational susceptibility occurs where association with another plant type increases an individual plant's risk of herbivory. This might be through close spatial association with individuals that are highly attractive to a mammalian herbivore whose mode of feeding does not allow selectivity at the scale of individual plants. It might be association with individuals of a species of plant that are highly susceptible to high insect infestation or a situation where insects colonize neighboring plants as they exhaust the food supply on one plant. Alternatively or additionally, it might be about association with a palatable species that is especially easy for herbivores to detect at a distance, thereby drawing more herbivores to the immediate vicinity of the focal plant of a different species. Conversely, it might be about association with an unpalatable species causing herbivores to focus more on palatable plants in their vicinity.

The study that first drew wide attention to associational effects was that of Hay (1986) on marine plants. This study also emphasized that there may be costs to associating with another species even if that species offers protection from predators. He found that palatable plants gained protection from herbivory by fish and urchins by growing in close association to a plant that these herbivores found aversive because the herbivores retreated after sampling an unpalatable plant. However, if the fish and urchins were experimentally excluded by caging, then palatable plants grew better on their own than when they faced competition for light and nutrients from a closely associated unpalatable plant. Thus the selective pressures on associational benefits in plants will be even further complicated by mechanisms other than herbivory.

Given the range of mechanisms discussed above, often a number will be in operation simultaneously, and thus whether species *A* benefits from association with species *B* will be defined by the relative balance of the strengths of different mechanisms, and thus in turn a function of intrinsic traits of the individuals of the species involved, their relative frequencies and spatial organization, traits of the herbivore concerned, and even the wider ecological setting and weather.

5.4 CASE STUDIES OF SELECTION PRESSURES ON MSGS

In this chapter and the preceding one, we have explored mechanisms that will influence fitness consequences of mixed-species grouping. Although we considered these mechanisms separately, we have emphasized that they will often occur in combination. We finish these two linked chapters by taking some of the most studied systems as case studies to explore what is known about the combined effects of different mechanisms.

5.4.1 Thomson's and Grant's Gazelles

Even when the benefits to both parties are similar, there can be complexity and asymmetry in the underlying mechanisms; and these may not fully be resolved in even the better discussed systems. For example, FitzGibbon (1990) reported on naturally occurring cheetah attacks on Thomson's and Grant's gazelles that often form both single-species groups and MSGs (FitzGibbon, 1990). Here both species appear to gain antipredatory benefits from mixed-species grouping. Firstly, when in a larger group (regardless of whether this is an MSG or single-species group), Thomson's gazelle benefitted because cheetahs tended to avoid attacking larger groups. Secondly, individuals in larger groups (regardless of composition) spent less time being vigilant and as a consequence spent more time for foraging. On top of these benefits, Thomson's gazelles gained further antipredatory benefits specifically from being in an MSG. These benefits were accrued because cheetahs were detected more reliably by groups that included Grant's gazelles; this may be because the Grant's gazelles are taller and have a more advantageous viewpoint with their head-up or because they tend to have their head-up more often than Thomson's gazelle. Cheetahs seek to stalk toward prey undetected before breaking cover and attacking. Once a cheetah has been detected by a group member (of either species), this gazelle shows behaviors (e.g., food stamping and directed staring) that indicate the

discovery both to group participants and to the approaching cheetah. Such discoveries often prompt cheetahs to give up their interest in that group without a chase. Those chases that did occur were less likely to be successful if the chase singled out a Thomson's gazelle from an MSG compared to a single-species group. FitzGibbon explained this last effect as the result of cheetahs breaking cover earlier during the stalking phase when approaching an MSG rather than a single-species group of Thomson's gazelle; this may be a good strategy in the light of enhanced risk of discovery by members of an MSG, particularly Grant's gazelles, during stalking. Set against all these benefits, there is a disadvantage to Thomson's individuals of being in an MSG. When a cheetah does launch a chase against a specific individual in an MSG, that individual is more likely to be a Thomson's. It may still pay Thomson's gazelle to join MSGs because of all the other benefits listed above, particularly if that species is fairly numerous within the MSG, so that the costs of cheetah preferences are diluted among many Thomson's. This in turn may suggest that MSGs between these species form by a single-species group of Thomson's fusing with a similar group of Grant's, rather than a single Thomson's joining a heterospecific group. FitzGibbon did not look as closely at the costs and benefits to Grant's gazelle beyond the clear advantage of grouping with another species that their shared predator prefers to attack. This has become such an iconic study that further investigation, in particular exploring the costs and benefits to Grant's gazelle, would be very valuable.

Before leaving this study, there are two further issues of note. Firstly, males more often participate in MSG than females. FitzGibbon explains this as being related to a male's interest in holding a territory and thus their intolerance of conspecific males; however, they have higher tolerance for males of the other species. Females have no similar drive to avoid females of the same species. Secondly, it is curious that although Grant's gazelles have superior vigilance against stalking cheetahs, Thomson's gazelles do not reduce their personal investment in vigilance any more when grouping with Grant's than conspecifics. FitzGibbon explained this as being because that Thomson's gazelle is also preyed upon by hunting dogs and spotted hyenas. These predators do not rely on stalking to get close to prey, and so the superior early warning abilities of the Grant's gazelle against stalking cryptic cheetah may be inconsequential against these other predator types.

5.4.2 Hornbills and Mongooses

Another iconic case of mixed-species grouping is the association between two hornbill species and dwarf mongoose, studied by Rasa (1983). Here the MSG seems to be a mutualism where mongooses gain protection from predators and hornbills gain food. Mongooses in a group move through the environment digging and searching for arthropod prey. Some of their prey (such as grasshoppers) jump into the air to flee the mongooses; it is generally these prey that are visible to and consumed by the hornbills that walk along the ground following the mongooses. It is assumed that this is beneficial to the hornbills although the net energy gain from foraging in this way versus foraging without mongooses has not been studied. Likewise, it is not clear to what extent (if any) the presence of the hornbills reduces the foraging efficiency of mongooses although it is assumed that any such cost is outweighed by the antipredatory benefits to mongooses of attendant hornbills. Hornbills emit alarm calls in response to approaching predators that mongooses respond to with appropriate antipredatory behavior (e.g., seeking cover). More interesting yet, Rasa showed that hornbills produced alarm calls in response to approach by both predators that were a threat to both species and to approach by predators that were a threat to mongooses but not to hornbills. Hornbills did not alarm call in response to approach of other large animals that were not a predatory treat to either species. Rasa also presents evidence that both species value the association. Mongooses sleep overnight in a termite mound, whereas the hornbills roost in trees, often some distance away. In the morning the hornbills fly to the overnight site of the mongooses; if the mongooses are not awake, then the hornbills begin loud calling that serves to rouse the mongooses, which soon after begin foraging. Conversely, if the mongooses awaken before the hornbills' arrival, then they wait for the hornbills before starting foraging. This is an interesting study system given the considerable behavioral adaptations to mixed-species grouping adopted by both parties. However, again further exploration would be welcome to fully characterize the costs and benefits to both parties.

5.4.3 Oxpeckers and Mammalian Grazers

Two African birds, the red-billed oxpecker and the yellow-billed oxpecker, find their food entirely on the coats of a wide range of ungulate mammals. Because the ungulates often associate, so do the birds; this can be thought of as a moving aggregation for them. Until the last two decades, this association was generally viewed as mutualistic—with the oxpeckers removing and consuming ectoparasites from the coats of the ungulates. However, recent research has highlighted the complexity in the relationship, with oxpeckers not only consuming ectoparasites but also consuming blood from the mammals, sometimes reopening healed wounds to access this blood (key recent publications are Weeks, 2000; Plantan et al., 2013; Bishop and Bishop, 2014). It seems likely that this association veers between mutualism and parasitism according to circumstances, with a given ungulate being more likely to suffer parasitism when it has a low ectoparasite load, has already-opened or recently-healed sores, lesions, or wounds, and/or if it is attractive relative to other potential hosts of the oxpeckers (Nunn et al., 2011). Another source of evidence that there is such a shifting balance between costs and benefits is that sometimes ungulates appear to solicit the attention of oxpeckers, but at other times they seem to behave so as to discourage the birds' attention (Bishop and Bishop, 2014).

This is another potential model system that would benefit from greater study on the costs and benefits to both partners. Specifically, the potential benefits to the ungulate are unclear. Although it seems self-evident that removal of ectoparasites by the birds should be beneficial to the ungulates, there are few demonstrations that the cleaning activity actually leads to lower parasite loads. Other potential benefits have been postulated but not explored in detail, such as reduced time spent in grooming and warning of the approach of mammalian predators by means of alarm calling by the birds (Bishop and Bishop, 2014). Whether the balance of costs and benefits experienced by ungulates is similar when attended by one or the other oxpecker species has also not been explored.

The costs of birds opening up wounds and drinking blood seem self-evident, but again there has been no attempt to correlate the attention from birds with the extent of exterior damage or some other measure of cost to the ungulates. Weeks (2000) excluded oxpeckers from access to some confined cattle but did not find that this led to increased tick burdens relative to control cattle. The costs of blood loss also include extending the time that wounds are open, which may increase the risk of infection or increase the attraction of insects that further irritate the host and exacerbate the wound. Costs of wound-feeding will be very dependent on the type of wound, the health of the ungulate, and the extent of wound-feeding by oxpeckers; but these costs too have not been explored. There is currently no evidence that ungulates with higher ectoparasite loads are either most likely to solicit or less likely to resist oxpeckers. From the oxpecker's perspective, detailed studies of captive individuals (Plantan et al., 2013) suggest that birds have a preference for blood over parasites but will generally choose to consume a mixture of the two in-cafeteria experiments. They are less likely to wound-feed when on hosts with a high density of preferred ectoparasites; but wound-feeding still occurred on such hosts. There is also evidence that oxpeckers preferentially feed from areas of hosts that contain higher ectoparasite densities. Yet, factors relating to the selection of one host over another and any impact of competition between birds are unclear (Nunn et al., 2011; Bishop and Bishop, 2014), although there is some evidence that oxpeckers prefer host species that typically have high parasite load (Fig. 5.5; Nunn et al., 2011).

5.4.4 Cleaner Fish

Some fish (and indeed other aquatic organisms) gain their food from the skin of larger fish. This has much in common with the case of oxpeckers above; originally it was considered that this was a mutualism with cleaners removing and consuming ectoparasites, but it is now clear that they can also feed on the mucous that coats the exterior of fish too. At least some types of cleaners can have a preference for mucous over parasites, but the extent of such mucous consumption can be controlled by the larger host fish, which can break away from the interaction with the cleaner, be less likely to use that cleaner again, or even attack the cleaner (Bshary et al., 2008). This interaction can be thought of as

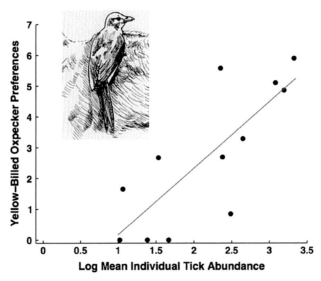

FIGURE 5.5 Oxpeckers focus on sites where ticks are abundant on hosts. The association between tick abundance on hosts at a given site and a preference measure that increases with the number of oxpeckers per host at that site. *Adapted from Nunn, C.L., Ezenwa, V.O., Arnold, C., Koenig, W.D., 2011. Mutualism or parasitism? using a phylogenetic approach to characterize the oxpecker-ungulate relationship. Evolution 65, 1297–1304, provided by Charles Nunn; sketch by Isabella Hoskins.*

an MSA, especially because cleaners often operate simultaneously on the same client. Indeed, mucous consumption is less likely during simultaneous cleaning because cleaners police their partners to avoid hosts prematurely breaking away from the cleaning station (Bshary et al., 2008). Furthermore, hosts often arrive at a cleaning station simultaneously, with observation of the cleaning of others being a key determinant of their behavior in subsequent cleaning interactions (Pinto et al., 2011). Specifically, when cleaners are observed by other potential clients, they are less likely to consume mucous. In addition, hosts that see a cleaner eating mucous from another are less likely to then submit to being cleaned by that individual.

Unlike the oxpeckers, there is good evidence that cleaner fish can reduce the burden of parasites on their hosts (Grutter, 1999). Indirect evidence for this also comes from the behavior of the hosts: they actively solicit the association with cleaners by visiting sites where cleaners are, by remaining still and opening their gills to allow access to the cleaners, and by avoiding preying on cleaners. Opportunity for exploitation of hosts may be less in this system than with oxpeckers because hosts have to submit to being cleaned and thus presumably only visit cleaners when they have ectoparasites, often have a choice of cleaners and are known to exercise that choice, and can break off an interaction and punish cheating cleaners.

5.4.5 Birds That Use False Alarm Calls to Kleptoparasitize

A number of bird species have been reported as foraging in an MSG and emitting alarm calls when no predator is present to induce a bird of another species to drop a particularly choice prey item as it flees; this item is then being consumed by the false alarm caller (Munn, 1986). The fact that other species do not simply ignore these calls suggests that the kleptoparasite also emits some reliable alarm calls; so in this system the kleptoparasite gains food rewards from participation in an MSG and the other species pay a cost in lost food but get a benefit in enhanced protection from predation. This idea has been more fully studied in the African fork-tailed drongo. This bird can either forage itself for small flying insects or join an MSG with other birds or small mammals and use either aggression or false alarm calls to access larger prey items that the other species generally dig out of the substrate. These appear to be at least partly mutually exclusive foraging options because kleptoparasitism requires being nearer the ground than is optimal for self-foraging. Drongos are shown to use kleptoparasitism by deception more when there are fewer flying insects, suggesting that kleptoparasitism can be used flexibly (Flower et al., 2013; Fig. 5.6).

Fork-tailed drongos commonly associate with groups of ground-feeding pied babblers. Perching while waiting for kleptoparasitic opportunities, the drongos emits a call that is considered to inform the babblers that it is acting as a sentinel, with babblers responding to this call by increasing their foraging efficiency (presumably because they invest less in personal vigilance, Radford et al., 2011). This increased foraging efficiency may benefit drongos and babblers by increasing kleptoparasitic opportunities. Perched drongos do emit alarm calls in response to terrestrial predators that are little threat to them when they associate with ground feeders in an MSG but not when self-feeding (Ridley et al., 2007), suggesting that alarm calling may require investment in vigilance by the drongos and brings an antipredatory benefit to the ground feeders. In a study involving drongos and ground-feeding social weavers, the weavers were drawn to sentinel-calling drongos and decreased vigilance and increased foraging success in response to these calls. The sentinel call was also used as an "all clear" signal immediately after weavers took flight in response to false alarm calls, causing weavers to return to foraging more quickly (thus mitigating the cost of false alarms and also benefiting drongos by enhancing the overall rate of kleptoparasitic opportunities, Baigrie et al., 2014). We will return to this example in Chapter 6, as part of the discussion of communication in MSGs.

5.4.6 Association Between Dolphins, Tuna, and Seabirds

This association is reliable enough to have long been exploited by fishermen as a way to find tuna, but we mention it briefly here mainly to emphasize that even for such a relatively well-known MSG identifying the costs and benefits involved rests mostly on supposition because of the challenges of data collection

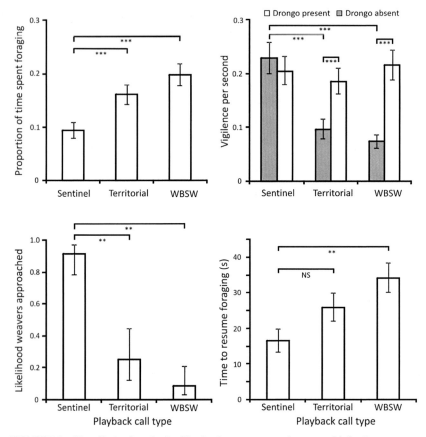

FIGURE 5.6 **The effects of sentinel calling by drongos on nearby weaverbirds.** In response to playback of drongo sentinel calls foraging weavers spent more time in searching for food (top left) and were less frequently vigilant (top right, *shaded bars*) than in response to control playbacks of drongo territorial calls or white-browed sparrow-weaver (WBSW) calls. When leaving their communal nests to forage, weavers were also more likely to approach playback of drongo sentinel calls than territorial calls or WBSW calls. Furthermore, after weavers had fled to cover in vegetation following playback of a drongo's false alarm call, they resumed foraging more quickly in response to playback of sentinel calls than WBSW calls, although there was not a significant difference from drongos' territorial calls. *Adapted from Flower, T.P., Child, M.F., Ridley, A.R., 2013. The ecological economics of kleptoparasitism: pay-offs from self-foraging versus kleptoparasitism. Journal of Animal Ecology 82, 245–255, and provided by Tom Flower.*

on these free-living oceanic animals. Tuna and dolphins have common predators (e.g., large sharks and orcas) and aggregating may bring antipredatory benefits. Mixed-species aggregating may bring improved detection of predators, with dolphins benefiting from tuna's olfactory sensing and tuna benefiting from dolphin echolocation. Dolphins corral fish into a ball and often drive that ball to the surface to minimize escape routes for the fish; this behavior may be exploited by both tuna and seabirds in their foraging (Scott et al., 2012).

5.5 CONCLUSIONS

We began this chapter by exploring potential antipredatory benefits of mixed-species grouping. It would seem that encounter-dilution is a benefit to grouping that should apply across a wide range of circumstances and taxa. We highlight plausible mechanisms by which there might be specific benefits to MSGs in this context and emphasize that both theoretical and empirical exploration of these should be fruitful. Shared vigilance and collective detection of predators are likely relatively commonplace in MSGs, and we also highlight examples where species in an MSG can combine complementary antipredatory detection abilities to offer particular benefits to an MSG. We suggest that the confusion and oddity effects deserve further empirical study, and the consequences of such effects (and nonrandom prey selectivity within groups more generally) for encounter-dilution benefits of MSG deserves further study. In some circumstances, MSGs will offer particular benefits through the ability of some species to deter would-be predators and through social learning about predators. We highlight the latter as being particularly in need of further study in a mixed-species context.

Many of the concepts by which mobile predators can influence selective pressures on mixed-species grouping in their prey are directly applicable to pressures exerted by mobile parasites on their hosts. There has been considerable investigation of this in single-species groups, and extension of this to MSGs would be highly appropriate. Even less studied are MSGs in the context of parasites and disease that are spread by contagion. MSGs have the potential to offer protection from such parasites and diseases, but this has been very underexplored empirically and theoretically. There are also parallels between mechanisms by which MSGs in animals offer protection from predators and those by which plants' risk of herbivory are influenced by close association with individuals of other species. However, the evolution of traits in this context is greatly complicated by the immobility of plants at almost all life-history stages, giving very limited scope for influence over their associations.

We concluded with examination of costs and benefits in some of the most studied systems. These examples highlight that even for well-known MSGs, we still have considerable empirical challenge to fully characterize the costs and benefits of these complex associations.

Chapter 6

Communication

6.1 INTRODUCTION

Without some means of communication between the various species in a mixed-species group (MSG), groups would not be able to form or maintain cohesion for long. Although communication between members of the same species is facilitated by the fact that all individuals produce and receive the same type of signals, communication between species faces the challenge of being intelligible to a range of species, each producing and receiving potentially different types of signals. Consider the case of the Guyana dolphin that changes the tone and pitch of its whistles when interacting with the distantly related bottlenose dolphin (May-Collado, 2010). Similarly, some species must learn the meaning of signals produced by other species to use them adaptively (Magrath et al., 2015b).

Interspecies communication encompasses a wide range of interactions between sender and receiver species, only some of which are relevant to MSGs (Fig. 6.1). At the simplest level, a species can inadvertently produce cues intended to no one or can send signals intended to conspecifics, which are then opportunistically intercepted by other species. Cues are like footprints on the beach that unintentionally betray the presence of a passerby earlier. Predatory bats, for instance, can attack frogs after detecting their mating calls, which are obviously intended for other frogs and not for their predator (Tuttle and Ryan, 1981). Interactions of this type are referred to as eavesdropping (Westrip and Bell, 2015). Signaling, by contrast, involves specially designed signals aimed at modifying the behavior of others (Bradbury and Vehrencamp, 2011).

Cues or signals produced by one species can be intercepted opportunistically by another without any further interaction between the species (Section 2.2). This is probably the case for ground-foraging skinks that eavesdrop on the alarm calls of red-vented bulbuls, a bird species foraging in the trees and with which no further interaction is possible (Fuong et al., 2014). For our purposes, eavesdropping is only relevant when the receiver species uses this information to form MSGs or maintain the cohesion of such groups. As an example, solitary scimitarbills eavesdrop on the alarm calls made by pied babblers but also seek their company to forage in MSGs (Ridley et al., 2014). In the following, we thus concentrate on species in the community that extract information from cues or signals produced by other species and that have the potential to form MSGs.

Mixed-Species Groups of Animals. http://dx.doi.org/10.1016/B978-0-12-805355-3.00006-3
Copyright © 2017 Elsevier Inc. All rights reserved.

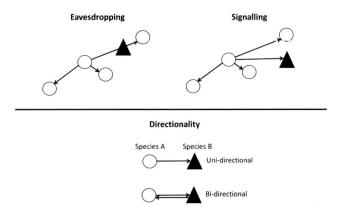

FIGURE 6.1 Framework for interspecies communication. Communication between species can involve eavesdropping or signaling (top panel). Different species are represented schematically by different geometric shapes. With eavesdropping, cues or signals provided by one species are intercepted by nonintended receivers of another species. Eavesdropping can be uni- or bidirectional depending on whether only one or both species can eavesdrop on the cues or signals provided by the other (bottom panel). Signaling involves signals that are intended to modify the behavior of another species. Such signaling can also be uni- or bidirectional.

There are costs and benefits associated with the production of cues or signals and also with the reception of these cues and signals. Such considerations are important to ultimately understand why cues or signals are produced by one species and why this information can be valuable to other species. The attraction of predators to frog-mating calls is a good example of a cost associated with the production of a signal, which must be taken into account when thinking about the evolution of calls to attract potential mates.

Receiver species can also experience costs and benefits after using cues or signals produced by another species. Some signals, for instance, can manipulate the receiver species into providing benefits for the signaling species at its own expense. We are all familiar with Aesop's fable about the boy who cried wolf. By crying wolf, the boy tricked villagers into thinking that a wolf was attacking their sheep. A similar ploy has been uncovered in many avian MSGs. In such groups, some species produce false alarm calls to scare other species away from food (Munn, 1986; Ridley and Raihani, 2007; Flower et al., 2014). Signals need not always be costly to the receiver species, and, in some cases, sender and receiver species can both benefit. Communication of this type can be unidirectional if only one species involved produces signals (Fig. 6.1) (Bshary et al., 2006). In the most complex case, communication is bidirectional and mutually beneficial to all concerned. In tropical forests of South America, different species of monkeys in the same community all produce calls that are intelligible to the other species and that allow the species to benefit from the formation of larger MSGs (Windfelder, 2001).

In the following, we explore the various types of communication that exist between species in an MSG from the simplest form of interceptive communication known as eavesdropping to the more complex cases involving honest or sometimes deceptive signals.

6.2 EAVESDROPPING

Eavesdropping is the simplest form of communication between species known to form MSGs. Eavesdropping can be unidirectional, in which case only one of the interacting species eavesdrop on the signals of the other. This is the case in the scimitarbill-babbler groups because babblers do not respond to the alarm calls made by scimitarbills while the reverse occurs (Ridley et al., 2014). In bidirectional eavesdropping, the different species involved can all produce and respond to such cues or signals. Eavesdropping can occur in a variety of contexts involving the location of resources, the location of groups, and the assessment of predation risk. Eavesdropping can also involve different communication channels from visual to vocal and chemical.

6.2.1 Eavesdropping to Locate Resources

Eavesdropping to locate resources can involve cues provided inadvertently by one species, which are used by nonintended receivers of another species. In other cases, receiver species can intercept signals produced by one species that are intended for conspecifics. In this section, we restrict our discussion to cases in which the production of such cues or signals is not detrimental to the sender species. Therefore, we leave aside cases involving prey species unwittingly providing cues about their presence to predators (Tuttle and Ryan, 1981) or species providing cues that can be used by heterospecific competitors to locate resources (Nieh et al., 2004; Dawson and Chittka, 2012). In such cases, there could actually be selection pressure to make such cues or signals less salient to potential predators and competitors.

Eavesdropping to locate resources is also known as local enhancement (Thorpe, 1956; Section 4.2.1). Local enhancement is well established between individuals of the same species (Galef and Giraldeau, 2001) but can apply just as well to cases involving individuals from different species. The production of inadvertent cues during foraging or breeding activities can hardly be avoided. In fish, for example, the simple act of feeding by one species can act a visual recruitment cue to other nearby species (Pereira et al., 2012). Similarly, the mud plumes produced by bottom-feeding gray whales attract any seabird within visual range (Obst and Hunt, 1990). If breeding sites are considered resources, conspicuous nesting activities of one species can be used as a cue by other species to settle in the same habitat (Seppänen et al., 2007).

In most cases, eavesdropping interactions between species are unidirectional in the sense that only one species acts as a provider of information about resource location, and other species benefit from gaining this information. In some aggregations, this results in asymmetric benefits: for instance, some species such as petrels arrive late into seabirds mixed-species associations, having cued off other species (Hoffman et al., 1981). Numbering in the hundreds or thousands, they then dive en masse into the school of fish, decreasing food availability for all other flock members. Nevertheless, there is also evidence of bidirectionality in certain associations of marine seabirds in which one species can act both as provider of information about the location of food and receiver of information from other species (Hoffman et al., 1981).

An aggregation of species around resources might form in response to an intercepted cue or simply in response to the actual presence of resources (Stamps, 1988), which would rule out eavesdropping. The best evidence for eavesdropping to locate resources across species comes from experiments that manipulate putative recruitment cues independently of resource location. A classic example involves heron MSGs in tropical lagoons (Caldwell, 1981). Aggregations of herons could simply arise in these lagoons because some areas consistently provide better foraging opportunities or the presence of feeding birds anywhere is attractive. A field experiment using realistic models of feeding herons showed that passing herons of different species landed near the models and actually preferred to land near snowy egret models rather than great egret models (Caldwell, 1981). Discrimination between model flocks showed that nearby herons did not simply land where any heron was feeding but actually preferred one species over the other (Fig. 6.2). Feeding snowy egrets, by

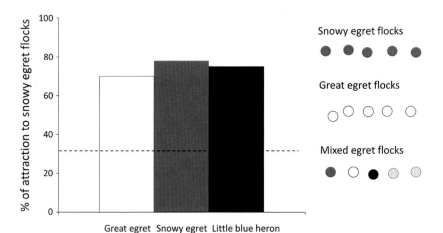

FIGURE 6.2 Attraction to mixed-species groups in herons. Various species of tropical herons, including white snowy egrets and great egrets and dark little blue herons, landed more often near experimental flocks of snowy egrets than to flocks of great egrets or flocks composed of a mixture of various species. Experimental flocks consisted of five decoys in a feeding posture. The *dashed line* shows the null hypothesis of random joining. *Adapted from Caldwell, G.S., 1981. Attraction to tropical mixed-species heron flocks: proximate mechanism and consequences. Behavioral Ecology and Sociobiology 8, 99–103.*

contrast to other heron species in the area, disturb fish from shallow waters, which then become available to other species. Landing near snowy egrets thus led to an increase in food intake rate. Great egrets tended to be more aggressive and rarely attracted other species. Eavesdropping here was thus unidirectional. Other experimental studies in birds also showed that the presence of feeding individuals from one species can affect the foraging choices of individuals of other species (Krebs, 1973; Waite and Grubb, 1988), but not always (Green and Leberg, 2005).

Use of visual cues to locate resources is also known in fish. Three-spined and nine-spined sticklebacks live in similar habitats and often form MSGs (Ward et al., 2003). In the laboratory, nine-spined sticklebacks showed a preference for the side of the tank where three-spined sticklebacks fed from a rich rather than a poor source of food, which indicates that cues provided by the feeding fish could not only help detect the location of food sources but also their relative richness (Coolen et al., 2003). Interestingly, three-spined sticklebacks showed no such preferences when provided with cues from feeding nine-spined sticklebacks, showing another example of unidirectional eavesdropping (Fig. 6.3).

In addition to visual cues, vocal cues can also be intercepted by other species to locate resources. Calls made by feeding gulls can be heard from a large distance and act as a catalyst for the formation of MSGs with other seabirds

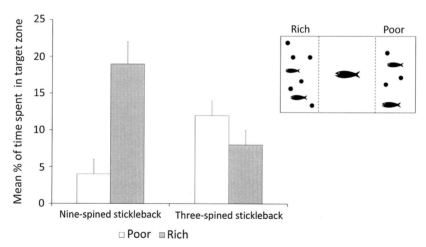

FIGURE 6.3 Unidirectional communication about the location of resources in fish. Individual nine-spined sticklebacks spent more time on average on the side of a tank where three-spined sticklebacks fed from a rich rather than a poor source of food, but such a preference was not exhibited by three-spined sticklebacks when exposed to demonstrator nine-spined sticklebacks. The inset shows a schematic version of the experiment in which a focal fish of one species faced a choice between two sides of the tank where individuals of the other species fed from a rich or a poor source of food. Means and standard error bars are shown. *Adapted from Coolen, I., van Bergen, Y., Day, R.L., Laland, K.N., 2003. Species difference in adaptive use of public information in sticklebacks. Proceedings of the National Academy of Sciences of the United States of America 270, 2413–2419.*

(Hoffman et al., 1981). Similarly, calls from ant-following bird species attract other species also known to exploit such resources (Chaves-Campos, 2003). Food-associated calls made by feeding monkeys can attract species that forage on their leftovers (Koda, 2012). As another example, willow tits that have uncovered resources produce long-distance calls that can attract not only conspecifics but also individuals of other species that have intercepted their calls (Suzuki, 2012).

In experimental as well as in observational studies of eavesdropping, lack of heterospecific attraction should not be interpreted as failure to eavesdrop. Cues or signals can be intercepted by one species, but some factors could actually prevent joining. For instance, subordinate species might be reluctant to join the food discoveries of a more dominant species (Waite and Grubb, 1988). In assessing eavesdropping, it is therefore important to consider the costs as well as the benefits of using the information provided by other species. In the stickleback example, the better-protected three-spined sticklebacks might be in a position to ignore cues provided by other species to locate resources because exposure to predators is less costly to them and foraging separately allows them to independently gather information about resources (Coolen et al., 2003).

In addition to visual cues, which were prominent in the above examples with birds and fish, chemical cues produced by one species can also be intercepted by other species to locate resources. This is particularly evident in fishes. Consuming prey and assimilating water-borne chemical cues from their environment, fishes exude chemical cues that can be used by conspecifics to locate resources (Ward et al., 2004). Recent work showed that individuals from other species can eavesdrop on these cues to locate food sources and, more generally, suitable foraging habitats (Webster et al., 2008).

6.2.2 Eavesdropping to Locate Groups

Eavesdropping can also be used to locate MSGs. Attraction to the location of food sources, as detailed in the previous section, immediately brings individuals from one species in contact with those of other species as well as the resources that they exploit. Eavesdropping to locate a group, therefore, can only be inferred in cases where it is possible to rule out a simple attraction to resources.

Convincing cases of eavesdropping to locate groups involve situations where a species eavesdrops on the cues or signals produced by another species but exploits different resources. Broadcasted calls of the greater racket-tailed drongo, for instance, are known to attract several species that form MSGs in tropical forests of Sri Lanka (Goodale and Kotagama, 2005b). Species that are attracted to such calls do not benefit from the resources exploited by drongos and, in fact, act as food flushers for drongos. Eavesdropping on drongo calls thus appears to be a means of forming groups in this system. Drongos can also modulate their calls to make them more attractive to other species (Goodale and Kotagama, 2006). Modulated calls probably represent signals rather than cues. Signaling is covered in Section 6.3.

Contact calls made by individuals of one species to maintain cohesion of their groups can be intercepted by other species to facilitate the formation of MSGs. In species of tits, long-distance calls can attract conspecifics as well as individuals from other species, but such calls tend to be made from rich feeding locations (Suzuki, 2012). In this case, it is not clear whether other species are attracted to the group or the resources. By contrast, long-distance calls in various species of tamarins in South America are often made when individuals are traveling or leaving their sleeping sites. Broadcasted long calls proved attractive to both conspecifics and nearby individuals of other species (Windfelder, 2001), suggesting that tamarins eavesdrop on the calls made by other species to attract conspecifics and use this information to form MSGs. The calls of these various species differed noticeably, which makes it unlikely that individuals were mistakenly attracted to the calls of another species. As we will discuss later on, long-distance calling might actually be a signal made by one species to specifically attract other species.

6.2.3 Eavesdropping to Mob Predators

The aforementioned cases of eavesdropping involved the location of resources or groups. Eavesdropping in such cases allows individuals of one species to increase their foraging efficiency or obtain general benefits associated with the formation of MSGs. Eavesdropping can also serve an antipredator function. In this case, cues and signals produced by one species facing a predator can be used by other nearby species to assess the motivation of that predator or the level of predation risk in general.

The detection of a predator typically sends prey species scurrying to safety. It would seem ill-advised to actually approach a predator because it would bring prey individuals closer to danger. Nevertheless, many species including birds, fish and mammals, are known to dice with danger by approaching a predator (Curio, 1978; FitzGibbon, 1994; Godin and Davis, 1995; Section 5.1.4). Approaches toward a predator by a potential prey species is known as predator inspection. After closing in, individuals can also mob the predator. Mobbing, which is often accompanied by loud calls, effectively advertises the position of a predator to all other prey species in the surroundings. The ensuing commotion often causes the predator to move away (Caro, 2005).

By contrast to alarm calls, which trigger evasive actions, mobbing calls attract conspecifics and heterospecifics alike to mount an aggressive response against the predator. Because of their loudness, mobbing calls are very good candidates for eavesdropping by other species (Marler, 1957). While any vulnerable species close enough can eavesdrop on mobbing calls, species that form MSGs probably have the most incentives to respond to such calls as they tend to cluster in space and face similar predators.

An experimental approach is needed to demonstrate eavesdropping on mobbing calls to rule out a simple reaction to the presence of a predator. Experiments

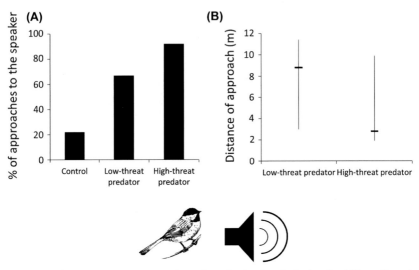

FIGURE 6.4 Eavesdropping on mobbing calls. After hearing playbacks of mobbing calls made by black-capped chickadees (inset), red-breasted nuthatches approached the sound source more often than after a control playback (A) and also approached closer when the mobbing calls indicated the presence of a more threatening predator (B). *Adapted from Templeton, C.N., Greene, E., 2007. Nuthatches eavesdrop on variations in heterospecific chickadee mobbing alarm calls. Proceedings of the National Academy of Sciences of the United States of America 104, 5479–5482.*

using mobbing call playbacks have been performed in a variety of species forming MSGs. One of the better-studied systems involves flocks formed around the black-capped chickadee, a small tit common in North American coniferous forests (Turcotte and Desrochers, 2002). This species produces alarm calls that contain information about predator type and level of danger (Templeton et al., 2005). In addition, chickadees also produce calls that elicit mobbing responses by conspecifics and heterospecifics alike. Broadcasted mobbing calls caused one such species, the red-breasted nuthatch, to approach the sound source rather than fleeing (Fig. 6.4). Moreover, the nuthatches also graded their approach according to the level of danger encoded in the mobbing calls (Templeton and Greene, 2007). Because chickadees and nuthatches have very different types of alarm calls, nuthatches probably need to learn the meaning of mobbing calls made by chickadees (Section 5.1.6). This learning is probably facilitated by the fact that the two species occur frequently together in MSGs (Morse, 1970). Nevertheless, conspecifics are probably more likely, in general, to be familiar with the subtle meaning of calls produced by one species than heterospecifics, which interact with a given species less often (Randler and Förschler, 2011).

As was the case with mobbing calls, distress calls can also attract other species. Such calls are made following capture by a predator. Potential functions of distress calls include startling the predator or attracting others to mob the predator. Broadcasted distress calls in one species of bird elicited mobbing responses by other species in the vicinity (Chu, 2001). Because distress calls in many species are quite similar in structure and can elicit responses even in nonsympatric species (Johnson et al., 2003), it is conceivable that distress calls might have evolved to explicitly attract other species. In the grip of a predator, any help from neighbors whether of the same species or not would be useful.

6.2.4 Eavesdropping to Assess Predation Risk

Eavesdropping on cues or signals that carry information about predation risk has attracted the most attention in the literature. Here again, we focus on interactions between species that can form MSGs. We thus leave aside cases of eavesdropping between species that do not interact any further (Dawson and Chittka, 2012; Goodale and Nieh, 2012).

Various cues or signals produced by one species can be used by another to assess predation risk. Most research focuses on cues or signals that betray a sudden increase in predation risk. An example of this is alarm calls produced by one species when a predator approaches. These calls are then used by nearby heterospecifics to initiate an escape response. In other cases, cues or signals indicate that predation risk is low or that an attack is not imminent. As an example, nonurgent contact calls produced by one species can be interpreted as an all-clear signal by heterospecifics (Radford et al., 2011).

In another situation, cues or signals betray past predation risk, which can be used to assess current predation risk in the habitat. In many aquatic species, breakage of the skin after an attack produces alarm substances that diffuse in the water and persist well after the predation attempt. Subsequent detection of these alarm substances can be used by conspecifics and heterospecifics alike to assess current danger and adopt appropriate responses such as forming tighter groups in some fish species (Brown et al., 1995). This interspecies alarm recognition system typically works with sympatric prey species that share a common predator. It is not clear in most cases whether this type of chemical communication is involved in the formation and cohesion of MSGs. However, forming such groups might be a way to learn about alarm substances produced by other species perhaps in a bidirectional fashion (Brown, 2003). In the following, we focus primarily on cues or signals that betray a sudden increase in predation risk, the most common situation.

Upon immediate predation risk, prey species can produce an array of reactions involving visual, vocal, or even vibrational elements. All of these cues can transfer information about imminent predation risk to nearby individuals. The literature on eavesdropping by other species on such alarm reactions has focused mostly on alarm calls, which can carry over large distances and tend to be very conspicuous. Nevertheless, visual cues of alarm can also be used by other species to assess predation risk. In fish and birds, visual fright reactions by one species in an MSG can trigger a similar fright reaction in other species in the group (Metcalfe, 1984; Krause, 1993b; Lima, 1995), a case of visual interspecies eavesdropping. When fright reactions involve a mix of visual and vocal components, it becomes more difficult to pinpoint the channel used to transfer information between species.

Alarm calls are produced by many animals in response to the presence of a predator. These calls are typically intended to conspecifics but can be intercepted by other species threatened by the same predator. Alarm calls can indicate not only that a predator is present but also the type of threat and the urgency of the threat (Seyfarth et al., 1980; Templeton et al., 2005). Using such information, other species can thus respond appropriately to the threat. A review of the literature reported evidence of eavesdropping on alarms calls in 87 species including birds, mammals, and reptiles (Magrath et al., 2015a). In many such cases, the species involved are not part of an MSG and one species simply eavesdrops on the alarm calls of another species with which it happens to share the same habitat and perhaps the same predators. An example of this involves ground-foraging lizards eavesdropping on the alarm calls of a tree-dwelling bird species mentioned earlier (Fuong et al., 2014; Section 6.1).

The procedure to determine whether eavesdropping takes place relies on responses to playbacks of alarm calls. Observational studies are more difficult to interpret because a response might be caused by direct detection of the predator rather than by eavesdropping on the alarm calls of another species. Eavesdropping can be inferred when an antipredator response is recorded following the playback of alarm calls but not after the broadcast of a control, non-alarm call, or after silence.

The experimental approach has uncovered several benefits but also constraints and perhaps even costs associated with eavesdropping on alarm calls produced by another species (Table 6.1). In an MSG context, eavesdropping on alarm calls is most useful for gaining information about immediate predation risk. Because of their spatial proximity, the various species in an MSG are all probably at high risk of attack when a predator approaches, which makes a response to alarm calls from another species all the more beneficial. Upon hearing the alarm call, a species can initiate a proper antipredator response based on the information available. Alarm calls from one species can produce a wide range of reactions in another species, including an immediate fleeing response, freezing (Sullivan, 1984; Hetrick and Sieving, 2011), an increase in antipredator vigilance (Ridley et al., 2010), or the production of alarm calls (Zuberbühler, 2000).

In some cases, alarm calls also contain information about the type of threats or the degree of danger. There is some evidence that species in an MSG can use such information to adjust their antipredator responses. Tufted titmice produce alarm calls that elicit fleeing or produce mobbing calls, which cause a gathering response near a predator that is not in attack mode. Carolina chickadees freeze when they hear the alarm call but approach and call when they hear the mobbing call (Hetrick and Sieving, 2011), suggesting an ability to discriminate between

TABLE 6.1 Benefits and Costs of Eavesdropping on Alarm Calls in Mixed-Species Groups

Type	Explanation	References
Benefits		
Decreased predation risk	Immediate detection of predators, of predator type or level of danger	Sullivan (1984) and Hetrick and Sieving (2011)
More accurate information about predation risk	More individuals are involved; greater breadth of coverage	Rasa (1983), Bshary and Noë (1997), Heymann and Buchanan-Smith (2000), and Goodale and Kotagama (2008)
Reduced vigilance	In response to better detection of threats	Bell et al. (2009), Radford et al. (2011), and Baigrie et al. (2014)
Increased foraging efficiency	In response to better detection of threats	Bell et al. (2009), Baigrie et al. (2014), and Ridley et al. (2014)
Costs or Constraints		
Learning costs	Learning might be required to recognize alarm calls	Wheatcroft and Price (2013) and Magrath et al. (2015b)
Low detection	Biased senses might limit the detection of alarm calls	Magrath et al. (2015a)
Low value	Alarm calls might not be relevant to eavesdropper, might provide wrong information, or might not be sensitive	Goodale and Kotagama (2005a,b, 2008) and Goodale et al. (2010)
Deception	Alarm calls might be used to usurp resources	Munn (1986) and Flower et al. (2014)

the two types of calls produced by titmice. In Africa, Diana monkeys responded to playbacks of alarm calls by Campbell's monkeys given to leopard or eagle predators as if the real predator was present (Zuberbühler, 2000), suggesting again a discrimination between alarm call types (Fig. 6.5).

The ability to eavesdrop on alarm calls can also allow individuals to determine that the situation is safe if no such calling takes place. For instance, pied babblers are known to eavesdrop on the alarm calls of fork-tailed drongos (Section 5.4.5). When hearing nonalarm calls from drongos,

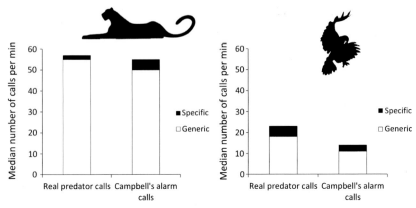

FIGURE 6.5   Mutual recognition of alarm calls in two species of monkeys. Following broadcasted alarm calls by Campbell's monkeys to two different types of threats (leopard on the left and eagle on the right), Diana monkeys adjusted their call rate as if the real predator that elicited the call by Campbell's monkeys was present and also matched the type of calls to the type of predator. The proportion of calls matching the predator type is shown in black in each bar. *Adapted from Zuberbühler, K., 2000. Interspecies semantic communication in two forest primates. Proceedings of the Royal Society of London B: Biological Sciences 267, 713–718.*

babblers and other species reduce their vigilance against predators (Ridley and Raihani, 2007; Bell et al., 2009; Radford et al., 2011; Baigrie et al., 2014).

The quality of information obtained from heterospecific rather than conspecific alarm calls is potentially of higher value. Relying on conspecifics, as well as heterospecifics, increases the number of individuals in a group that can be counted on to provide information about predation risk (Goodale and Kotagama, 2005a). Perhaps more importantly, the breadth of coverage of the habitat for potential sources of danger is wider because different species in an MSG often occupy different areas and might be more responsive to different types of threats. For mammals feeding close to the ground, eavesdropping on the alarm calls of birds in the trees can provide information about predators that they cannot readily see (Rasa, 1983; Sharpe et al., 2010). Diana monkeys in tropical forests of Africa forage high in the trees in large groups, making them a valuable source of information for other monkey species foraging below (Bshary and Noë, 1997).

Given all these benefits, why would some species ignore alarm calls produced by other species? Eavesdropping requires an ability to extract information from calls produced by another species. Lack of response could simply indicate a failure to detect such calls because the signal is too degraded or outside the range of typical communication. It seems quite unlikely that signal degradation can account for lack of response in an MSG where all species are in close proximity. Species in an MSG are often closely related phylogenetically and often use similar modes of alarm calling (Wheatcroft and Price, 2013), making it unlikely that calls are outside their typical communication range. Nevertheless, when communication involves distantly related species, such as mammals and birds, it is conceivable that some alarm calls might not be detected by other species because of sensory biases.

If detection is possible, lack of response might indicate an inability to recognize the call as an alarm call. This is especially relevant for heterospecific alarm calls whose recognition involves learning (Wheatcroft and Price, 2013; Magrath et al., 2015b). In this case, some species might not be able to respond to alarm calls from other species because their learning is biased toward conspecific calls (Madden et al., 2005). By contrast, some species with particularly well-developed intraspecific alarm calling might be predisposed to learn other species' alarm calls (Lea et al., 2008). The context in which learning takes place could also have an influence. Situations in which individuals can experience repeated exposures to an association between alarm calls from another species and a particular predator without risking their lives are probably more conducive to proper learning. Animals in stressful situations learning more quickly might also be the case (Kelley and Healy, 2012).

Beyond such constraints on using information from heterospecific alarms calls, lack of response could simply indicate that such calls have intrinsically low value (Table 6.1). Alarm calls from other species could be irrelevant if they signal the presence of predators that pose little threat or if they tend to be unreliable or insensitive. Indeed, it would make little sense to respond to alarm calls from a species that often fails to distinguish real from imagined threats or that is less able to detect predators (Magrath et al., 2009; Goodale et al., 2010; Westrip and Bell, 2015).

A final consideration is that some of these alarm calls could provide deliberately false information (Section 6.3.2). For interspecific eavesdropping to evolve, the costs of responding to such calls must be lower than the benefits. Deceptive communication falls into the concept of signaling because the aim is to modify the behavior of others. Signaling is covered in the following section.

6.3 SIGNALING

Interspecific signaling differs from eavesdropping in that the information provided by one species aims to modify the behavior of a receiver species. Signaling can be difficult to distinguish from eavesdropping because it is not always clear to whom the cues or signals are directed. Recording a reaction by one species to cues or signals produced by a sender species need not indicate that heterospecifics were specifically targeted. In the following, we specify under which conditions signaling between species can be inferred. We make a distinction between cases where the receiver species benefits from the information received (honest signals) and those where the receiver species is manipulated by the sender species and experiences a cost (deceptive signals).

6.3.1 Honest Signals

Honest signals implies benefits for the species that receives the signal. There are actually only a few cases of mutualistic interspecies communication involving species in an MSG. In all these cases, eavesdropping could be safely ruled out because heterospecifics were the only possible audience.

The strongest evidence comes from systems in which no conspecifics are nearby when a signal is produced. As described in Section 4.2.5, single groupers in the Red Sea make a vigorous body-shaking motion that entices giant moray eels to join them for coordinated hunts in the reefs (Bshary et al., 2006). Such signals recruit eels, and not groupers, and are only made when groupers are hungry. Similar recruitment signals aimed at other species have also been reported in another fish species (Lönnstedt et al., 2014). Vocalizations made by honeyguides to guide humans to food sources also fall in this category because these signals are clearly not intended for other honeyguides (Isack and Reyer, 1989).

Another situation involves alarm calls that are specifically modulated to reflect the need of the targeted species. As we saw in the preceding chapter (Section 5.4.2), hornbills that are associated with terrestrial mongooses produce alarm calls in response to the presence of raptors that prey on mongooses but not on them (Rasa, 1983), which strongly suggests that the calls were intended for the mongooses and not for other hornbills. Similar audience effects have also been noted in other avian species (Ridley et al., 2007).

A more contentious case of interspecies communication involves long calls of two species of tamarins in the tropical forests of South America. Groups of emperor and saddleback tamarins often form stable associations over extended periods of time. Long calls in tamarins carry over large distances and act as contact calls or territorial calls. Long calls were broadcasted in situations where groups of the two species were not in contact. The two species of tamarins produced long calls in response to broadcasted long calls of not only their own species but also to broadcasted long calls of the other species (Windfelder, 2001). The evidence clearly shows that different species of tamarins can recognize each other's long calls, and this has also been noted in other primate systems (Shultz et al., 2003), but it is not clear whether the long calls are produced to specifically attract the other species. Such calls might be made to defend the territory against nearby conspecific groups or to attract scattered conspecific group members.

Another way to rule out eavesdropping consists in showing that the putative interspecies signals are modified from their basic form to facilitate communication with other species. One possible example of this is Guyana dolphins that change the structure of their whistles when interacting with bottlenose dolphins (May-Collado, 2010).

6.3.2 Deceptive Signals

Deceptive signals are produced by one species to manipulate other species into providing benefits to the sender species at their own expense. In most cases, signaling can be easily inferred because the species that produces deceptive signals has no conspecific audience.

Deceptive communication in MSGs is particularly evident in avian groups. The first convincing evidence for deceptive signals in avian MSGs was provided by Munn who studied alarm calling in western Amazonia. Some species in these large groups, which can contain dozens of species (Section 3.5.6), act as an early warning system for the rest of the group. These species are typically sallying predators, a foraging mode that allows individuals to monitor their surroundings for danger while searching for passing prey. Upon spotting danger, these species emit loud alarm calls that send the rest of the group scurrying to safety. The plot thickens because they also produce alarm calls when no predator is present. That these calls are deceptive can be inferred from the fact that the callers benefit from the rush to safety by capturing prey left behind (Munn, 1986). These calls are thus an avian equivalent of crying wolf. Playbacks of deceptive calls trigger the same alarming response in the group, suggesting that honest and deceptive alarm calls have similar properties; they are, however, produced for sharply different reasons.

A similar strategy has also been uncovered more recently in another study system in South Africa (Section 5.4.5). Fork-tailed drongos in these flocks produce both honest and deceptive alarm calls that are identical in structure. Alarm calls cause other species in the group, including other avian species and also mammals, to abandon food to thieving drongos (Ridley et al., 2007; Ridley and Child, 2009; Flower, 2011; Radford et al., 2011; Flower et al., 2014). False alarm calls probably work because of the high cost of ignoring such calls by the receiver species. Ignoring an alarm call can increase food intake rate, but this benefit hardly compensates for failing to flee when a real predator is fast approaching. Used sparingly, such calls can be trusted most of the times to provide accurate information and are hard to ignore because it is impossible to distinguish real from deceptive calls.

6.3.3 Mimicry

A final case for interspecies communication involves mimicry. Vocal and visual mimicry in an MSG falls into the category of signals that evolved to modify the behavior of other species in the group. Mimicry signals can be honest or deceptive (Ruxton et al., 2004).

Drongos can not only acquire resources deceptively by producing their own false alarm calls, but they can also mimic the alarm calls of other species to the same effect. Vocal mimicry of these alarm calls is just as effective at enticing other species in the group to abandon their food (Flower et al., 2014). This strategy might have evolved to reduce habituation by other species in the group to deceptive drongo-specific alarm calls. By mimicking other species' calls, drongos make it increasingly difficult for other species to ignore any alarm calls.

Another species of drongo can mimic the contact calls of other species. Such vocalizations proved attractive to other species (Goodale and Kotagama, 2006). This species can also mimic predator vocalizations, which proved quite effective in eliciting a fleeing response in other species in MSGs (Goodale et al., 2014). Vocal mimicry of mobbing calls in this species is also used to elicit mobbing responses in other nearby species (Goodale et al., 2014). These results show that other species in the group can readily discriminate between different types of calls and can just as easily be manipulated by the drongos. The fact that vocal mimicry is a flexible trait under the immediate control of the individual producing the calls makes it very likely that these calls are signals to other species.

Visual mimicry, by contrast to vocal mimicry, is typically a passive signal that cannot be dynamically controlled by the species producing the signal. Visual mimicry represents convergence in the visual appearance of different species involved in a MSG. Visual mimicry is known in birds and fish, and might have evolved for different reasons in these two groups.

In birds, visual mimicry has been linked to at least three different contributing factors. Mimicry could facilitate cohesion among group members because of the economy of signals that have to be learned to coordinate different species in a moving group (Moynihan, 1968). Mimicry in avian groups has also been linked to dominance because a subordinate species might benefit by resembling a dominant species to reduce interspecific aggression while feeding together (Diamond, 1982; Prum, 2014). In these two cases, the target of mimicry is another species in the MSG. Mimicry has also been linked to a reduction of the oddity effect, which is important because species that look different from others in the group could be disproportionately targeted by a predator (Barnard, 1979; Landeau and Terborgh, 1986). With the oddity effect (Section 5.1.3), the target of mimicry is a species outside the MSG, the predator, but the MSG provides the context in which visual mimicry evolved.

Two alternative hypotheses must be ruled out to provide a compelling case for visual mimicry. Firstly, because cases of mimicry in an MSG involve species living in the same habitat, resemblance can simply represent independent evolution of similar traits in response to shared habitat requirements (Willis, 1976; Burtt and Gatz, 1982). Secondly, because species in an MSG are often drawn from the same families, closely related species in such groups might have inherited similar traits from a common ancestor (Fig. 6.6). Mimicry studies also face the challenge of measuring the degree of convergence in appearance among the various species involved. Visual mimicry targets the eyes of other species in the group or the eyes of a predator, and yet most cases of mimicry have been judged with our own human eyes. Avian vision, in particular, is quite distinctive from human vision because birds have different color receptors in their eyes (Cuthill, 2006). Nevertheless, while we probably miss subtleties available to avian eyes, our perception of color is fortunately considered a good approximation of avian vision (Seddon et al., 2010).

Model
Semipalmated sandpiper

Putative mimic
Western sandpiper

Ancestry control
Little stint

FIGURE 6.6   **Visual mimicry in avian groups.** Semipalmated sandpipers and western sandpipers often forage together in groups on their wintering grounds in South America. The two species look remarkably similar, providing a potential case of visual mimicry. However, the Western Palearctic little stint, a sister species to both species of sandpipers, has a similar appearance despite living on a different continent. This strongly suggests that visual resemblance in all these species reflects a plumage pattern inherited from a shared ancestor. *Adapted from Beauchamp, G., Goodale, E., 2011. Plumage mimicry in avian mixed-species flocks: more or less than meets the eye? Auk 128, 487–496. Photo credits: Semipalmated sandpiper (Guy Beauchamp), western sandpiper (P.E. Hart) and little stint (Ron Knight).*

Two of us (Guy Beauchamp and Eben Goodale) recently reevaluated the evidence for visual mimicry in avian MSGs in the light of the alternative hypotheses mentioned above (Beauchamp and Goodale, 2011). In many cases of putative mimicry, we found that simpler explanations involving either inherited patterns from a common ancestor or shared habitat requirements could account for visual resemblance between species in MSGs. In other cases, mimicry seemed a reasonable possibility, but the evidence would be stronger if the factors that promoted mimicry could be determined. While our eyes judged the similarity in appearance between species quite compelling, it remains an issue whether the intended targets of mimicry would also perceive this similarity to the same extent.

Cases of visual mimicry in MSGs are also known in fish. A particularly intriguing case involves species that mimic and join another species to form MSGs. However, in such groups, the mimicking species can get closer to unsuspecting prey of yet another species. The mimicry works because the mimicked species is not a threat to the prey species, which reduces wariness in the prey species (Sazima, 2002). Aggressive mimicry of this type is quite common in fish (up to 39 species according to Sazima, 2002) and is also known in birds (Willis, 1963). The signal carried by aggressive mimicry is intended for potential prey outside the group. While the mimicking species obviously benefits from the association with another species, the mimicked species probably does not. In some cases, the mimicry could even be costly to the mimicked species if the prey species became aggressive to all members of the group in response to attacks by the mimicking species. In this case, it would be to the advantage of the mimicked species to discourage the mimicking species from joining it and to the advantage of the mimicking species to look even more identical. As always in cases of mimicry, alternative hypotheses related to habitat constraints and shared phylogenetic history should always be considered when evaluating the evidence. The fact that mimicry can be facultative in many of these fish species, which can change their color depending on the context, rules out these alternative hypotheses (Côté and Cheney, 2005).

Other cases of mimicry in fish are more straightforward and are akin to mimicry in avian groups (Dafni and Diamant, 1984). For instance, MSGs of grunts and goatfishes are common in reefs off the Brazilian coast. These two fish species closely resemble one another, at least as far as our eyes can perceive it. Resemblance in this case is thought to increase group cohesion and to make it harder for predators to zero in on a prey in a relatively large and homogenous group (Krajewski et al., 2004). In this situation, the mimicry benefits both species, but the rarer species in these associations has probably more to lose by not resembling the more numerous species. In another case, an octopus changed its coloration and posture to match those of another species during joint foraging bouts (Krajewski et al., 2009), a strong indication that mimicry was not accidental.

6.4 CONCLUSIONS

The surge of interest in eavesdropping and signaling between species (see Fig. 1.2) places studies of MSGs at the forefront of this research because strong interactions between species are needed for the formation and cohesion of such groups. The hard question is always the same in such studies: are species really communicating with one another, in the sense that signals are produced to modify the behavior of other species, or are species simply intercepting cues aimed at no one or at conspecifics?

The debate is not new. Long ago, Marler, in his studies of alarm and mobbing calls in different species of forest birds, noted similarity in the structure of these calls in a range of species sharing the same habitat (Marler, 1957). While one could argue that calls have converged to facilitate communication between species, as the economy of signals hypothesis would suggest, such calls might also be similar because they evolved in the same habitat and serve the same purpose of either hiding the presence of the signaler to the predator or facilitating recruitment of others to mob the predator. It is thus conceivable that these signals are aimed at the predator or other conspecifics and are intercepted opportunistically by other species that just happen to use similar channels of communication.

The case for interspecies communication is stronger in systems where signals produced by one species are modulated according to the need of the targeted species. This was the case for flexible vocal calls made by drongos and facultative mimicry in fish. In other cases, it remains important to determine whether conspecifics represent the true target of the cues or signals and how shared habitat requirements and phylogenetic inertia can account for interspecies communication.

Our discussion focused mostly on foraging MSGs, but the same type of research could also be carried when animals are breeding or resting (Chapter 2). Mixed-species aggregations are common during the reproductive season in birds (Burger, 1981) and mammals (Goldsworthy et al., 1999; Berthier et al., 2006) and also occur when animals are resting (Eiserer, 1984). Breeding activities, just like foraging, provide conspicuous cues that different species could use for eavesdropping (Nuechterlein, 1981; Burger, 1984). Information acquired from eavesdropping could be particularly useful when searching for a suitable breeding location (Farine et al., 2014). Future studies could confirm whether breeding activities by one species attract other species to the same site. As mentioned in Section 2.3.1, this has already been shown for species that breed in the same neighborhood but not for species that settle as a group at the same breeding location (Mönkkönen and Forsman, 2002).

Chapter 7

Leadership and Sentinel Behavior

7.1 INTRODUCTION

In any social organization, the question comes naturally as to whether different members of the group play different roles. Army units and management teams, to take two well-known cases in human groups, are clearly composed of individuals playing strictly defined roles. In fact, the effectiveness of the unit or team often is critically dependent on clearly defined role boundaries for each group member. The unit or team typically has a task to perform, and roles are assigned by a third party to achieve this goal. Lacking immediate goals and with no external role assignment, can we expect to see similar roles in mixed-species groups (MSGs)?

To assess this issue, it is important to make a distinction between fixed and flexible roles. We refer to fixed roles in an MSG as those that follow from particular attributes of a species, attributes that are apparent regardless of whether the species is inside or outside of an MSG, or, if inside such a group, what is the composition of that group. The occurrence of fixed roles reflects the simple fact that different species in an MSG are not clones of one another and can show substantial variation when it comes to searching for resources or allocating time to vigilance. Reflecting this variation, some species are more attractive to other species when it comes to forming an MSG. One example of a fixed role in an MSG is food flushing. A particularly dynamic species in such groups can disturb prey during its foraging making them available to nearby species (Belt, 1874; Struhsaker, 1981; Sazima et al., 2007; Sections 3.6.1 and 4.2.2). The role of food flusher was not devised to increase the effectiveness of the group in generating resources but reflects rather the natural foraging behavior of some species, which remains in the same shape and form regardless of whether other species are present. Similarly, some species in MSGs act as sentinels for predator detection (Munn, 1986; Ragusa-Netto, 2002). The sentinel role is not assigned by decree to optimize predator detection at the group level. Instead, this role might reflect the natural tendency of a species to forage in the higher stratum of the habitat or to use a foraging technique that allows individuals to concurrently monitor the surroundings (Martínez and Zenil, 2013).

Mixed-Species Groups of Animals. http://dx.doi.org/10.1016/B978-0-12-805355-3.00007-5
Copyright © 2017 Elsevier Inc. All rights reserved.

A flexible role, by contrast, is one that occurs specifically in the MSG context and that provides benefits to the species playing that role and, sometimes, incidentally, to other species in the group. This would be the case if a species, say, only acted as a sentinel in the presence of other species or switched foraging mode in an MSG to flush prey. Obviously, such roles have not evolved or are not performed to ensure the better functioning of the MSG, as would be the case for a unit or a team with a specific task to perform. When thinking about flexible roles, it is conceivable that some species evolved traits or can modify existing traits to make them more suitable to play certain roles in such groups. Traits associated with flexible roles should allow individuals that possess them to increase their own success when participating in an MSG.

In this chapter, we examine the occurrence of roles in MSGs and whether there is any evidence that roles are flexible rather than fixed. Two roles in particular have attracted much attention in the context of MSGs: leadership and sentinel behavior. Leadership is believed to facilitate the formation and cohesion of MSGs. Sentinel species act as an early warning system to detect predators. Much of the evidence for roles in MSGs comes from the avian literature, but the concept of roles also applies to other types of species such as fish (Strand, 1988; Sazima et al., 2007) and primates (Cords, 1990; Bshary and Noë, 1997; Smith et al., 2003).

7.2 LEADERSHIP

7.2.1 Historical Perspective

Early naturalists clearly noted that different species in MSGs play different roles. For instance, in his book detailing his trip to the Amazonian jungle, Henry Walter Bates recalled how the indigenous people thought that a little gray bird fascinated the rest of the group and led them across the jungle (Bates, 1863). Not surprisingly, no such species could ever be located, but clearly the question as to whether one species in an MSG plays a leadership role attracted attention early in the literature.

Avian MSGs that form in the more open habitats of Australia also appeared to coalesce around a few species of thornbills, which were labeled "association formers," whereas other species were characterized as "joiners" (Gannon, 1934) (Table 7.1). Gannon also suggested that calls from thornbills attracted other nearby species. Also in Australia, Hindwood (1937) noted that one species tended to initiate most movements in MSGs. In deciduous forests of Europe, the incessant movement of tit species was believed to attract other nearby species (Delamain, 1933).

In view of such differences among species in the ability to lead groups, Winterbottom (1943) distinguished between nucleus and circumference species to describe species that formed the core of MSGs in African woodlands and those that only joined groups occasionally and tended to occur at the periphery, respectively. The nucleus species typically foraged in monospecific groups

TABLE 7.1 A Lexicon of Terms Used to Describe Leadership Roles in MSGs

Terms	Definition	References
Association former	A species that facilitates the formation of MSGs	Gannon (1934)
Association joiner	A species that tends to join other species in MSGs	Gannon (1934)
Nucleus species	A species found at the core of MSGs	Winterbottom (1943)
Circumference species	A species found at the periphery of MSGs and typically less frequent than the nucleus species	Winterbottom (1943)
Accidental species	A species more rarely encountered in MSGs	Davis (1946)
Nuclear species	A species that facilitates the formation and cohesion of an MSG	Moynihan (1962)
Regular attendant species	A species that occurs regularly in an MSG but shows no leadership quality	Moynihan (1962)
Long-term follower	A species that follows an MSG over an extended period of time (similar to regular attendant species)	Buskirk et al. (1972)
Occasional attendant species	A species that occurs occasionally in an MSG but shows no leadership quality	Moynihan (1962)
Short-term follower	A species that follows an MSG for a short period of time and over short distances (similar to occasional attendant species)	Buskirk et al. (1972)
Active species	A species that more often joins other species than it is joined by others	Moynihan (1962)
Passive species	A species that is more often joined by other species than it joins	Moynihan (1962)
True leader species	A species that is followed by others much more than it joins others	Morse (1970)
True follower species	A species that follows others much more than it is joined	Morse (1970)
Sentinel species	A species that maintains high vigilance against predators from a vantage point	Westcott (1969), Conner et al. (1975), Munn (1986)

MSGs, mixed-species groups.

and also occurred frequently outside MSGs. Circumference species were also labeled as accidental in descriptions of avian MSGs in Brazil a few years later (Davis, 1946). Davis also identified one particularly vocal species commonly found in most groups and who seemed to act as a leader. These early descriptions clearly suggest that different species occur to different extent in MSGs and that some traits, such as calls or energetic movements, probably facilitate the formation and cohesion of these groups.

The classification of roles in MSGs took a more prominent turn after the publication of a review on the evolution of MSGs in Neotropical birds (Moynihan, 1962). In his classification scheme, Moynihan recognized nuclear and attendant species (Table 7.1). Nuclear species form the core of the groups, and these species are believed to be important for the formation and cohesion of these groups. Attendant species can be found regularly or more occasionally in these groups and play a more minor role in leadership. Moynihan also distinguished between active and passive species. An active species tends to join more often than it is joined by others, whereas a passive species tends to be joined more often than it itself joins. The passive, nuclear species would thus be the most important species in the formation of MSGs and might possess or evolved traits that facilitate grouping. Moynihan further speculated that such leaders possessed specific traits such as many contact calls, nonaggressive interactions, and dull plumage. Influential reviews on MSGs still refer to the scheme proposed by Moynihan minus the distinction between active and passive species (Diamond, 1981; Powell, 1985; Terborgh, 1990; Hutto, 1994; Greenberg, 2000). How to practically distinguish between different types of attendant species and how regularly they attend MSGs relative to their abundance outside such groups also represent an active field of research (Farley et al., 2008).

7.2.2 How to Quantify Leadership

The obvious question about leadership is how to distinguish between leader and follower species. Moynihan provided a list of qualitative traits for classifying species into different roles. However, without a quantitative way of classifying species as leaders or followers, it is difficult to move beyond subjective impressions. Various ways have been used over the years to assess leadership in MSGs in a more quantitative fashion, and these are reviewed below.

We restrict our discussion to MSGs involving more than two species. The reason for this is simple: in groups with two species, leadership is often totally confounded with foraging mode. A food-flushing species leads by default because the other species must follow it to obtain the flushed prey. As an example, due to their large size and their relative wasteful feeding habits, many primate species flush or drop food for other species, including birds and mammals (Heymann and Hsia, 2015). In nearly all such cases, the nonprimate species follows the primate species. Relationships of this kind are also common in coral reefs in which a variety of fish that disturb the surface of the reef, releasing food particles, are followed by other species (Sazima et al., 2007). Although obligate following of this kind can also occur to some extent in groups involving more than two species, it is probably less likely because foraging modes are more diverse and the benefits of joining a group with more species are not just limited to flushed prey. Another potential confounding factor in two-species groups is dominance. The species that leads might be the one that has access to resources first due to higher dominance, as seems to be the case in some primate MSGs (Heymann and Buchanan-Smith, 2000). Again, strict dominance is less likely to affect leadership in groups with more species. For similar reasons, we also exclude mixed-species associations that aggregate around resources. In this sort of groups, some species uncover resources that then become available to other nearby species (Chapter 2). Both dominance and foraging mode typically determine which species uncover resources and those that subsequently follow.

7.2.2.1 Quantitative Assessment

The classification scheme proposed by Moynihan is purely qualitative and cannot really be used practically to identify roles in MSGs. When using this scheme, most researchers simply look for traits that are presumed to be important for the formation and cohesion of groups (Section 7.2.3). A species should also occur frequently in MSGs to at least qualify as nuclear. The classification of a species as a leader or as a follower too often remains a qualitative exercise fraught with ambiguity (Willis, 1972).

Morse (1970) on avian MSGs in temperate forests of North America developed a more quantitative approach, which addressed some of these shortcomings. Morse proposed two indices for each species in an MSG: (1) the ratio of the number of species led to the number of species followed and (2) the ratio of the number of times individuals of the species led to the number of times they followed (Fig. 7.1). Depending on the values of these ratios, Morse distinguished between true leader and true follower species, and those in between. A true leader species often leads and is frequently followed by several species, whereas the reverse is true for a true follower. True leaders would be characterized as passive nuclear in Moynihan's scheme, whereas true follower species would represent the active attendant species in that scheme.

Morse's approach is limited to cases where it is possible to unambiguously determine when one individual leads and another follows. Morse determined a leader and a follower in interactions between individuals of two particular species based on the latency to respond to a movement initiated by one individual. Morse did not specify how closely in time a joining event should follow the initial movement. What is also missing is a sense of the direction taken by the follower species after the initial movement by the leader species. Presumably the movement must be made in a similar direction, but again it was not clear how far would a follower have to stray before the movement was discounted as a following event.

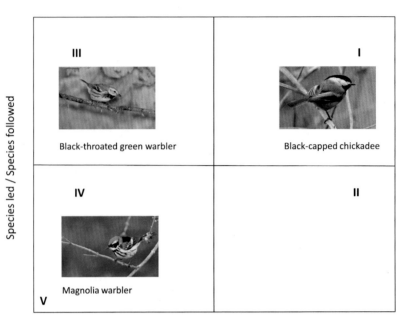

FIGURE 7.1 Leadership in mixed-species groups (MSGs). Classification scheme proposed by Morse (1970) to assess leadership in MSGs. Species are characterized by two ratios: the ratio of the number of species led to the number of species followed, and the ratio of the number of times individuals of a species led to the number of times they followed. True leader species can be found in the upper right quadrant (I), whereas true follower species can be found in the lower left corner (V). In Maine woods, the black-capped chickadee led most MSGs; the black-throated green warbler, in quadrant III, led fewer times but attracted more species than the magnolia warbler in quadrant IV. No species fitted in quadrant II. *Photo credits: Black-capped chickadee (Brendan Lally), black-throated green warbler (Bill Majoros), and magnolia warbler (Isaac Sanchez).*

Perhaps the greatest difficulty with the ratio approach proposed by Morse is that it fails to take into account the frequency with which different species interact in MSGs. Two different species might have the same ratio of the number of times the species led to the number of times it followed even though one species could lead and join many more times than the other. In short, the use of ratios cannot distinguish between a species that is followed by many species when it leads but who leads infrequently and those that are equally attractive to other species but lead more frequently. Similarly, ratios are independent of the amount of time spent in an MSG. A case for a leader species would be stronger if for the same ratio the leader and follower species interacted over several hours as opposed to minutes. Time spent interacting was indeed proposed as a way to classify interactions in fish MSGs (Pereira et al., 2012), implying a need to consider how long species remain together while foraging. The ratio approach proposed by Morse has rarely been adopted by other researchers. Nevertheless, the general idea of using data on which species follow others is simple and useful to get a first sense of leadership in an MSG.

Another approach to quantifying leadership consists in correlating the movement of a group across the habitat to the movement of an alleged leader species. The movement direction adopted by the group is expected to match the movement direction taken by the leader species rather than by follower species. Close matching of this kind was observed in MSGs including the tufted titmouse, a presumed leader species (Contreras and Sieving, 2011). This approach is best suited to less obstructed habitats in which the direction of movement can easily be determined over long distances.

7.2.2.2 Experimental Manipulation

If a species plays an important role in the formation and cohesion of MSGs, MSGs without a leader species should be less likely to form or persist. Observational evidence suggested that in areas where leader species are absent, MSGs are less frequent and show a different organization (Stouffer and Bierregaard, 1995; Maldonado-Coelho and Marini, 2004; Zhang et al., 2013). Because absence of the leader species might reflect factors that also affect follower species, experimental evidence is needed to isolate the effect of leader species on MSGs. An experimental approach holds much promise to avoid the pitfalls associated with observational studies of leadership.

In one of the few experiments carried out to investigate whether leader species really are involved in the formation and cohesion of MSGs, researchers removed putative leader species from woodlots in eastern North America and documented subsequent adjustments by the remaining species. Earlier observations suggested that the tufted titmouse and the Carolina chickadee lead these MSGs. Various species of woodpeckers and nuthatches typically coalesce near these species during foraging. In woodlots where leader species were removed (and relocated elsewhere), close spatial associations between various follower species occurred less frequently than in woodlots where the leader species remained (Fig. 7.2), showing the importance of leader species for the formation of MSGs in these woodlots (Dolby and Grubb, 1999). This finding should not be interpreted as evidence that leader species actively recruited other species to form groups in these woodlots. Passive attraction would be sufficient to form MSGs in this case. We will return to this study once more in discussing the conservation implications of MSGs (Section 8.4).

Experiments have also focused on putative signals made by leader species that facilitate the formation of MSGs. Such signals can be visual or vocal (Section 7.2.3). If visual signals are important to attract other species, masking such signals should affect the formation of MSGs. If vocal signals play an important role, playback of such signals should attract other species. Several studies used an experimental approach to examine these predictions.

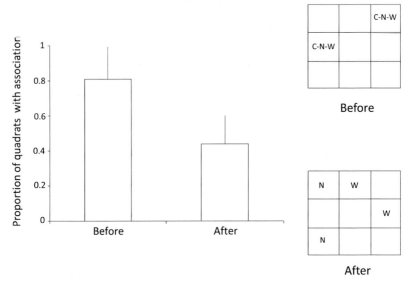

FIGURE 7.2 Formation of mixed-species groups. The proportion of quadrats in a woodlot in which follower species occurred together decreased after the leader species were relocated elsewhere. In the schematic representation of the experiment on the right panel, the letter C refers to chickadees, the leader species, and the letters W (woodpeckers) and N (nuthatches) refer to follower species. Means and standard error bars are shown. *Adapted from Dolby, A.S., Grubb, T.C., 1999. Functional roles in mixed-species foraging flocks: a field manipulation. Auk 116, 557–559.*

The white-flanked antwren is a gregarious bird commonly found in the tropical forests of Central and South America. This species has long been thought to act as a leader of large understory MSGs (Moynihan, 1962; Wiley, 1971; Willis, 1972; Jones, 1977; Munn and Terborgh, 1979). In this species, wing-flashing by foraging males exposes white flanks on an otherwise dark body. Conspicuous wing-flashing might thus work as a highly visible recruitment signal for nearby species in the dark understory (Wiley, 1971). Contrary to expectation, dyeing the white flanks of males with black paint influenced neither the size of MSGs nor the duration of flocking (Botero, 2002). Despite negative results, this experimental approach is laudable because it puts to the test the idea that certain signals favor the formation and cohesion of MSGs.

If the presence of other species brings foraging benefits to the leader species, satiated leaders should be less inclined to produce recruitment signals. In the case of interspecific coordinated hunting in two fish species of the Red Sea discussed before (Sections 4.2.5 and 6.3.1), groupers initiated joint hunting forays by vigorous visual signals aimed at eels (Bshary et al., 2006). The production of such signals would be a good example of a flexible role in an MSG as groupers only behave in such a way in the presence of eels. Experimentally fed leaders produced no such signals, indicating that the purpose of the signals is most likely linked to the attraction of the other species. Experimental manipulations

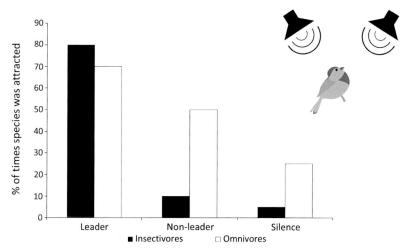

FIGURE 7.3 Attraction to vocal signals from leader species. Playback experiments in a Sri Lankan rainforest revealed that non-leader bird species were attracted more often to the calls of putative leader species than to those of a non-leader species or silence. The attraction was stronger for insectivorous species than for omnivorous/frugivorous species. *Adapted from Goodale, E., Kotagama, S.W., 2005b. Testing the roles of species in mixed-species bird flocks of a Sri Lankan rainforest. Journal of Tropical Ecology 21, 669–676.*

of this sort are interesting because they directly address the putative value of signals made by leader species.

The experimental approach has also targeted vocal signals made by putative leader species. The principle behind this approach is simple: broadcasted calls or vocalizations made by a putative leader species should elicit approaches or interest from species known to associate with them in MSGs. This is the approach used by one of us (Eben Goodale) to determine which of the two species suspected to act as leaders was more attractive to other species in Sri Lankan MSGs (Goodale and Kotagama, 2005b). Nearby foraging species tended to approach the broadcast source more often after hearing calls made by the two leader species than those made by a non-leader species or following silence, which strongly suggests that calls alone can attract other species and be a facilitating factor in the formation of MSGs (Fig. 7.3). Interestingly, it was the combination of calls from the two putative leader species rather than calls from either species alone that proved the most attractive. Other experimental studies also showed that vocal signals can attract nearby species in bird MSGs (Mönkkönen et al., 1996; Goodale et al., 2012; Suzuki, 2012; Cordeiro et al., 2015) and also in monkey MSGs (Cords, 2000; Windfelder, 2001).

The above studies are useful to determine whether vocal signals can attract other species. They do not address what happens after the species have aggregated. After aggregation, other signals alongside vocal cues could play an important role in maintaining cohesion. It is also important to consider the possibility that calls from another species elicit interest because they are perceived as a challenge to defended resources rather than as a signal to form an MSG. This hypothesis is more easily dismissed when joining species are not territorial. Overall, the main advantage of playback experiments is that attraction to another species can be separated from attraction to particular food patches, which could be associated with the presence of that species.

The playback approach can also be used to investigate whether calls made by leader species possess features that are especially attractive to other species. In the Sri Lankan rainforest groups mentioned earlier, the racket-tailed drongo, a leader species, benefits from the presence of food-flushing species. Drongos are also known to mimic calls from other species. It turns out that playbacks of drongo calls with more mimicked elements proved more attractive to other species (Goodale and Kotagama, 2006). Calls with more mimicked elements might give the impression that more species are present in the habitat. Whether mimicry has evolved to play that purpose is not known (Section 6.3.3).

7.2.2.3 Null Models

Fitting species into specific roles in an MSG works best when a species almost always leads or follows. In less black and white cases, leadership can be variable and depend on a host of factors such as species composition (Diamond, 1987) or habitat type (Morse, 1970; Gram, 1998). Despite its shortcomings, the indices developed by Morse (Section 7.2.2.1) provided a way to measure leadership on a continuous scale. Leader species would occur at one end of the continuum and follower species at the other end. The challenge now is to develop a continuous measure of leadership that takes into account the availability of other species in the habitat. The use of null models, which can specify how often different species are expected to cooccur by chance in an MSG, allows us to classify a species according to its degree of association with other species. A leader species would be expected to have stronger ties with other species or to occur with other species more often than expected from a null model of random associations among species.

Recent work tackled this issue by using data from avian MSGs in lowland tropical forests of India (Srinivasan et al., 2010). The idea was to compare the observed pattern of cooccurrences between pairs of species to that expected under a null model of cooccurrences. A large positive deviation from the expected value would suggest that a species is more nuclear or plays a more central role in MSGs. In this tropical system, many species cooccurred at levels well above the null expectation, which suggests that particular species sought one another.

The number of interspecies associations above this null expectation can be used as a continuous measure of nuclearity, which can then be related to various ecological factors believed to be associated with nuclear species. In particular, the number of interspecies associations positively correlated with the intraspecific group size of a species (Fig. 7.4), which supports the idea that a nuclear species is more gregarious (Moynihan, 1962). For one type of flock in this system, the number of interspecies associations was also larger for sallying species than for gleaning species. Sallying species are often believed to act as sentinels in MSGs because they actively scan their surroundings for prey and can incidentally detect predators more easily (Wiley, 1971; Munn and Terborgh, 1979). This study thus provides support for the hypothesis that sallying species are attractive to other species although in fairness the association could run the other way if sallying species seek other species to obtain food.

One drawback of the methodology just described is that the nuclearity index is study-dependent: the number of interspecies associations must be a function of the number of available species, which might vary across study systems. This makes it difficult to compare the index across studies or ecological conditions. A further issue is that in systems with a large number of species, some species are likely to be closely related. As such, they might share similar tendencies to lead or follow, in which case their nuclearity index values might be quite similar and nonindependent. Inertia in nuclearity index values caused by relatedness would be an issue of statistical concern in such analyses. Finally, the nuclearity

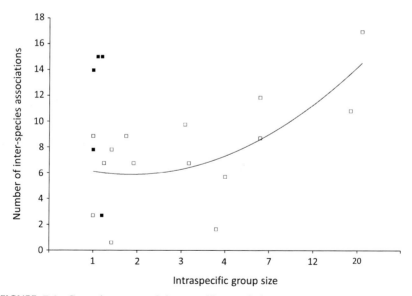

FIGURE 7.4 Gregariousness and interspecific association patterns. In understory avian mixed-species groups of lowland tropical forests in India, species that tend to associate more often with other species than expected by chance also tend to occur in larger groups. Sallying species in these flocks (*black squares*), which scan the surroundings for aerial prey, are distinguished from gleaning species (*open squares*). *Adapted from Srinivasan, U., Raza, R.H., Quader, S., 2010. The nuclear question: rethinking species importance in multi-species animal groups. Journal of Animal Ecology 79, 948–954.*

index does not address the issue of leadership per se, that is, whether a species actually leads others. It is quite clear that a species must cooccur more than expected by chance with other species to act as a leader.

Null models have also been used to address the issue of leadership more directly. In groups composed of two species, it is quite reasonable to expect that at least one of the species benefits from the association. A null model for the composition of groups composed of two species was formulated using the occurrence of gregarious and nongregarious species in tropical evergreen forests of India (Sridhar and Shanker, 2014a). As it turned out, gregarious species occurred disproportionately more often in two-species groups than expected by chance. Observations of such groups in the field also revealed that the gregarious species led more often than the others. Nuclear species in fish are also known to occur in intraspecific groups, and the approach developed here for birds could be used to determine whether associations including gregarious species of fish occur beyond the level expected by chance (Sazima et al., 2007).

Null models can be seen as a first step to identify associations between particular interacting species that occur above the level predicted by chance. Once nonrandom associations are identified, the purported benefits of such associations can be sought and perhaps related to specific roles in groups (Srinivasan and Quader, 2012). Spatial and temporal changes in the strength of these associations can also be used to infer benefits, as has been done recently in MSGs of large East African mammals (Kiffner et al., 2014).

7.2.2.4 Social Network Analysis

At its very core, MSGs represent a collection of individuals from at least two species that interact at a rate that exceeds what would be predicted by random encounters in the same habitat. If some species lead such groups, we would expect such species to be at the core of these groups, that is, they would be involved in many more associations with other species than predicted by chance. The null model approach presented earlier allows us to quantify such associations, but this method gives the same weight to all nonrandom associations by simply culling the number of associations above the random level. What we need is a tool to map out relationships between all individuals of all potentially interacting species in a specific area during a specific time period. Metrics from such a network of interacting individuals could be derived to determine not only the number but also the strength of interactions among individuals and species. Clusters of individuals from a particular species might become apparent in the network, which could be linked to particular roles in the group.

Social network analysis provides just this sort of tool (Croft et al., 2008). Social network analysis records associations or interactions between different individuals from different species over a fixed period of time in a particular habitat or location. The collective links between individual and species nodes constitute the social network. Metrics have been developed to characterize such networks, which allow us to get a quantitative estimate of the sociality of an individual and of the species in general in such a network. For instance, individuals or species that form the core of a group would be expected to have a high "centrality" value as they are involved in more associations. Such estimates can be compared for different species and across ecological conditions (Wey et al., 2008). Returning to the study where leader species were experimentally removed from woodlots (Dolby and Grubb, 1999), the social network analysis of species interactions before and after the manipulation would provide the quantitative estimates necessary to statistically compare the two situations.

Social network analysis can be applied to interactions between individuals from just one species (Lusseau, 2007) or to interactions between individuals of several species in the same habitat. As an example, the network of interactions between an introduced species of fish in a novel habitat and local species was drawn to highlight the potentially disturbing consequences of the invasion in terms of relationships between native species (Beyer et al., 2010). Social network analyses of interactions between species that can actually form groups together are of particular interest for our purpose.

Social network analysis carried out at the level of the individual requires individual identification. In many field situations involving MSGs, individual identification is not possible given the large pool of interacting individuals or their distance from the observer. In such cases, the social network analysis can be carried out at the species level (Sridhar and Shanker, 2014a). However, it is clear that any variation in association strength within species will be missed. This is important because we often think of a species in an MSG as having fixed attributes, while it is often the case that substantial variation can occur among individuals of the same species. For instance, in MSGs of tit species interacting at feeders, larger individuals tended to have more associates than smaller ones, perhaps reflecting the fact that subordinate birds try to avoid others (Farine et al., 2012). An analysis carried out at the species level would not be able to identify this trend, if dominance acts mostly within rather than among species.

Recent applications of social network analyses to MSGs have been able to shed new light on species roles in groups. In East Africa, large browsing mammals often congregate in MSGs (FitzGibbon, 1990; Burger and Gochfeld, 1994; Pays et al., 2014; Schmitt et al., 2014). Earlier studies focused on the costs and benefits that some of these species experience in MSGs (Chapters 4 and 5), but species differences in attractiveness to others and whether the structure of these groups varies across ecological conditions were not addressed until recently.

Abundance estimates of species occurring along stretches of roads in two ecosystems in East Africa allowed researchers to establish whether some species occurred more frequently than expected by chance in MSGs and whether some species played a more central role in these associations (Kiffner et al.,

2014). In particular, plains zebras tended to be involved in many more associations than expected by chance and showed strong patterns of cooccurrences with other species, especially wildebeest (Fig. 7.5). Associations weakened during the dry season, perhaps in response to a decrease in food availability, which would reduce the benefits of foraging closely with other species. Another study showed that zebras decreased their vigilance in the presence of wildebeest, a species that is preferred by their common predator, the African lion (Schmitt et al., 2014). Such benefits might explain why zebras and wildebeest are so strongly associated and suggest that zebras might seek other species to reduce predation risk.

Social network analyses have also focused on MSGs in other species including birds and even bats (Ancillotto et al., 2015). In thornbill flocks in Australia, the analysis revealed again the central role of putative leader species (Farine and Milburn, 2013). In addition, each individual was marked with a distinctive series of color leg bands, and it was possible to see that females had different social connections than did males. Other studies contrasted the structure identified by social network analysis between different geographical areas (Sridhar et al., 2013) and over several years and across environmental conditions (Anguita and Simeone, 2015).

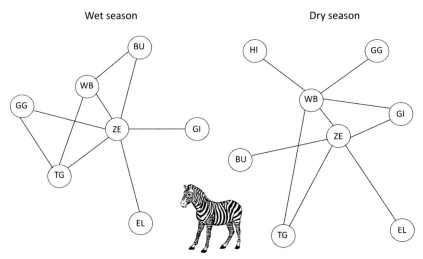

FIGURE 7.5 **Social network analysis of East African mixed-species groups (MSGs).** Browsing mammals in the Tarangire–Manyara savannah ecosystem in East Africa aggregate in MSGs in which plains zebra play a central role. The strength of the association between pairs of species is negatively proportional to the length of the bar relating each species. In the dry seasons, associations became noticeably weaker (*BU*, cape buffalo; *EL*, eland; *GG*, grant's gazelle; *GI*, giraffe; *HI*, hippopotamus; *TG*, Thomson's gazelle; *WB*, wildebeest; *ZE*, plains zebra). *Adapted from Kiffner, C., Kioko, J., Leweri, C., Krause, S., 2014. Seasonal patterns of mixed species groups in large East African mammals. PLoS One 9, e113446.*

As with null models, social network analysis cannot reveal which species is actively responsible for the formation and cohesion of MSGs. However, it provides a means to identify key species or individuals in such groups. Quantitative indices produced by social network analysis can be used to test hypotheses about the factors driving the formation of MSGs.

7.2.3 Characteristics of Leader Species

Several traits by leader species are believed to play an important role in the formation and cohesion of MSGs. In the following, we identify such traits in different taxa.

Following the publication of Moynihan's scheme to classify the role of species in MSGs, avian researchers routinely reported whether some species in their study system tended to be nuclear. Compiling findings from many study systems, they made it clear that in birds many characteristics occur frequently in leader species. Specifically, these species have a high propensity to occur in such groups, form intraspecific groups, and are often conspicuously active and vocal (Hutto, 1994). Level of activity is quite subjective and might be hard to assess empirically. But the relationship between leadership and gregariousness is much easier to determine by simply measuring intraspecific group size.

Why would gregariousness be an asset for leader species? A large group size could simply increase visual conspicuousness, making it easier for nearby species to join and follow the leading species over time and through space. Noisiness associated with frequent contact calls in a large group could also ease detection from afar. Other traits might also be associated with group size and further facilitate flocking. Conspicuous alarm calling, perhaps selected for when there are large groups of related individuals (Maynard Smith, 1965), might act as a cue to locate the leader species, rallying other species rapidly to maintain group cohesion (Goodale and Kotagama, 2005b).

In a survey of Neotropical avian groups, Powell reported that group size tended to be slightly larger in leader species than in those that typically follow (Powell, 1985). Two of us (Eben Goodale and Guy Beauchamp) extended this analysis to a much broader range of study systems worldwide (Goodale and Beauchamp, 2010). Avoiding circularity is important in such an endeavor because leadership is often defined by a large group size. We thus looked for evidence of leadership independent of group size. Leader species often occur at the forefront of groups and initiate more movements than the others. We might expect gregarious species to lead flocks simply because they outnumber other group members. In this sort of random model, it would be unlikely to consistently find all members of the leader species at the forefront of the group.

We contrasted mean flock size across different types of species in each study system: the leader species, one associate species (a species commonly found in such groups, but not a leader), and in one occasional species (occurring in less than 50% of the groups). In our worldwide survey, mean group size was indeed

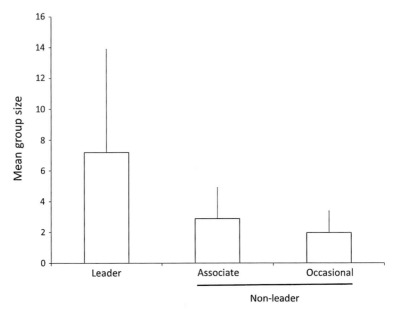

FIGURE 7.6 Leadership and intraspecific group size in avian mixed-species groups (MSGs). A worldwide survey of various MSGs in birds revealed that mean intraspecific group size was larger in species that led groups than in non-leader species, including those that are commonly found in groups but do not lead (associate) and those that occur less frequently in groups (occasional). Means and standard deviations are shown. *After Goodale, E., Beauchamp, G., 2010. The relationship between leadership and gregariousness in mixed-species bird flocks. Journal of Avian Biology 41, 99–103.*

significantly larger in leader species than in associate and occasional species from the same groups (Fig. 7.6). Such differences in intraspecific group sizes between nuclear and attendant species might also hold in fish MSGs, but this has not yet been analyzed rigorously (Lukoschek and McCormick, 2000).

In another worldwide survey of MSGs in birds, Sridhar et al. (2009) found that cooperative breeding was more common in leader than in follower species. Cooperative breeders often live in kin groups (Cockburn, 1998), so again we might expect well-developed communication and alarm signals in such groups. Such signals could be eavesdropped on by other species to increase their safety. However, further study is required to identify the exact traits and behaviors associated with cooperative breeding that actually help to promote MSGs in birds.

Leadership has also been addressed in different species of tamarins that often form MSGs in tropical forests of South America. Here, dominance status appears to influence leadership, as the smaller species in the mostly two-species groups formed by these monkeys typically leads (Terborgh, 1983; Peres, 1996; Smith et al., 2003). Leadership probably increases the chances of accessing resources before the more dominant species can monopolize food patches. Dominance status in MSGs also affects leadership in dolphins (Quérouil et al., 2008) and in fish (Lukoschek and McCormick, 2000; Sazima et al., 2007).

The production of calls appears to serve as a signal to facilitate the formation of MSGs in tamarins (Pook and Pook, 1982) as well as in many other species of primates (Cords, 2000). What is not known is whether the production of calls or the particular attributes of these calls are different in leader and non-leader species and whether these calls are specifically targeted at other species or are simply made to attract conspecifics and heterospecifics alike.

7.3 SENTINEL BEHAVIOR

In several species of birds and mammals, and even in fish, vigilance against predators can be coordinated at the group level (Bednekoff, 2015). In such groups, high vigilance maintained by one or a few individuals allows other group members to focus their attention on other activities (such as resting or foraging). This sort of vigilance is also often carried out from vantage points providing a better view of the surroundings. Coordination of vigilance at the group level and selection of vantage points to carry out vigilance characterize sentinel behavior (Bednekoff, 2015).

Sentinel behavior has been noted very early on by naturalists (Wallace, 1875; Cary, 1901; Elliott, 1913). In Chacma baboons, for instance, Elliott (1913) noted that one or a few individuals located on a prominent rock warned the rest of the foraging group about impending danger. Sentinel behavior has typically been investigated in single-species groups, but recent studies show that its occurrence in MSGs is not rare and might in fact be a force that favors the formation and cohesion of such groups.

Coordination of vigilance represents a hallmark of sentinel behavior in single-species groups. In a coordinated group, individuals take turns to perform vigilance; which implies that vigilance is maintained at a stable level over time. As far as we know, coordination of vigilance between species has not been documented in MSGs. In most cases, one species takes advantage of the higher vigilance provided by another species without reciprocating.

In an MSG, a sentinel role can be attributed to species that perform a disproportionate share of the vigilance and that can warn other species more rapidly or effectively than would be the case if the other species maintained vigilance on their own. For this purpose, sentinel species tend to provide frequent and conspicuous alarm calls. Based on these traits, sentinel species have been identified in many MSGs in birds and in mammals (Table 7.2).

High vigilance in a sentinel species has been related to several factors including foraging technique, vantage position, large individual size, and large group size. In terms of foraging technique, sallying species, as noted earlier, tend to scan their surroundings to a greater extent when searching for prey, which might incidentally facilitate the detection of predators. Large size might be useful to get a better view of the surroundings. A large group size increases the chances

TABLE 7.2 Cases of Sentinel Behavior in MSGs of Birds and Mammals

System	Sentinel Species	Attributes of the Sentinel Species	References
Birds			
North American scrubland	Pinyon jay	High position	Balda et al. (1972)
Peruvian understory	Bluish-slate antshrike	Sallying	Munn and Terborgh (1979)
Peruvian canopy	White-winged shrike-tanager	Sallying	Munn and Terborgh (1979)
European heathland	European stonechat		Greig-Smith (1981)
North American temperate forest	Black-capped chickadee		Sullivan (1984)
Brazilian savannah	White-banded tanager	High position	Alves and Cavalcanti (1996)
Brazilian understory	Cinereous antshrike		Stouffer and Bierregaard (1996)
Brazilian savannah	Chalk-browed mockingbird	Large group size	Ragusa-Netto (1997)
Brazilian understory	Black-goggled tanager	Sallying	Maldonado-Coelho and Marini (2000), Maldonado-Coelho and Durães (2003)
Brazilian savannah	White-rumped tanager	High position	Ragusa-Netto (2000)
Taiwanese forest	Grey-cheeked fulvetta	Large group size	Chen and Hsieh (2002)
Sri Lankan understory	Greater racket-tailed drongo	Sallying	Goodale and Kotagama (2005a)
South African savannah	Fork-tailed drongo	Sallying	Morgan et al. (2012), Baigrie et al. (2014)
Tanzanian forest	Square-tailed drongo	Sallying	Cordeiro et al. (2015)
Mammals			
African savannah	Grant's gazelle	Large size	FitzGibbon (1990)

TABLE 7.2 Cases of Sentinel Behavior in MSGs of Birds and Mammals—cont'd

System	Sentinel Species	Attributes of the Sentinel Species	References
Ivory Coast forest	Diana monkey	High position and large group size and spread	Bshary and Noë (1997)
South American forest	Moustached tamarin	High position	Heymann and Buchanan-Smith (2000)
Ivory Coast forest	Diana monkey	High position and large group size	Wolters and Zuberbühler (2003)
South African savannah	Meerkat	High position and large group size	Waterman and Roth (2007)
Birds–Mammals			
East African desert	Von der Dercken's and Eastern yellow-billed hornbills	High position	Rasa (1983)
South African savannah	Fork-tailed drongo	High position	Sharpe et al. (2010)

MSGs, mixed-species groups.

of detecting threats by adding more eyes and ears to the detection of predators. Similarly, vantage positions such as the top of trees can increase detection distance. Alarm calling might be more frequent in sentinel species because of higher vigilance, but other factors can predispose such species to be more vocal. As noted earlier, many sentinel species are also cooperative breeders, which are suspected to produce more frequent and elaborate alarm calls. In addition to frequent alarm calls, some sentinel species also produce calls to attract other species (Goodale and Kotagama, 2006; Baigrie et al., 2014; Goodale et al., 2014). Whether sentinel behavior in an MSG constitutes a flexible or fixed role is not clear. In many cases, the sentinel species appears to play this role involuntarily; it simply reflects its tendencies to produce more alarm calls or to occupy a better position for antipredator vigilance. For instance, the Diana monkey in tropical forests of Africa lives in large groups and typically forages high in the trees. This species is thus more likely to raise the alarm than other species foraging in smaller groups at lower strata (Bshary and Noë, 1997; Wolters and Zuberbühler, 2003). Alarm calling in this species appears aimed at conspecifics (or predators) (Zuberbühler et al., 1997), and there is no indication that alarm call features are any different when this species forages in MSGs. There is also no indication that different species in such groups coordinate their vigilance.

Perhaps the best evidence for a flexible role in sentinel species comes from MSGs where the sentinel species also forages alone, in which case particular features of sentinel calls or behavior can only be related to the occurrence of other species. Deceptive alarm calling and the use of calls to attract other species can be seen as features that evolved in the context of MSGs. For instance, when they forage alone, fork-tailed drongos, a bird species that often associates with other species in South African savannah (Section 5.4.5), give alarm calls after detecting aerial predators but tend to ignore terrestrial predators (Ridley et al., 2007). When joining another species that forages on the ground, the drongos also provide alarm calls after detecting terrestrial predators that cause a threat to the individuals below. Other work shows that drongos can mimic the calls of other species to attract other species to form MSGs and also to acquire resources deceptively (Goodale and Kotagama, 2006; Baigrie et al., 2014; Flower et al., 2014; Sections 6.3.2 and 6.3.3).

7.4 CONCLUSIONS

Roles in MSGs have been related to leadership and sentinel behavior. We have reviewed several procedures to identify leaders in MSGs. The use of more objective tools to assess leadership is a welcome development that will surely help to clarify the various roles played by different species. Leaders and sentinels have now been documented in many MSGs in fish, birds, and mammals.

In many cases, roles coopt particular attributes of a species with seemingly little adjustments to the multispecies context. In terms of leadership, for instance, we lack evidence that the larger groups of leader species evolved to attract other species more effectively. Large groups, instead, might simply reflect selective demands for foraging in single-species groups. Similarly, frequent and more elaborate calling by leader species could reflect evolution for the benefit of kin in cooperative breeders.

The best evidence for flexible roles in MSGs comes from studies in which leader or sentinel species use specific signals to attract or deceive other species. This was the case for groupers using a signal to entice eels to form hunting groups. This signal was only used when groupers were hungry and was not aimed at other groupers (Bshary et al., 2006). Alarm calls by sentinel species tailored to a multispecies context (Goodale and Kotagama, 2006; Ridley et al., 2007; Flower et al., 2014) or the use of deceptive alarm calls by sentinel species (Munn, 1986; Baigrie et al., 2014) also strongly suggest that sentinel species actively respond to the presence of other species in their groups.

Future work could focus on visual or vocal signals produced by leaders and sentinel species to determine if specific features of these signals facilitate the formation or cohesion of MSGs. In addition to group size, Moynihan predicted that other features of leader species could play a role in birds, including their coloration (Moynihan, 1960). Signal convergence in MSGs, whether learned or acquired through evolution, might also be expected to facilitate communication in such groups (Diamond, 1981; May-Collado, 2010; Beauchamp and Goodale, 2011).

Chapter 8

Mixed-Species Groups and Conservation

8.1 INTRODUCTION: THE NEED FOR CONSERVATION OF MIXED-SPECIES ASSOCIATIONS

It is said that people only value something fully when they lose it. If this is true, we had better hope that people realize their loss quickly. For by many measures we have entered a sixth mass extinction crisis (Barnosky et al., 2011), although this time the driving agents are human induced and not naturally occurring. Indeed, there is now a push toward renaming our current geologic epoch the "Anthropocene" to recognize that anthropogenic change to natural ecosystems (hereafter referred to as "human disturbance") is now "a major geological and environmental force, as important as, or more important than, natural forces" (Corlett, 2015). The size of our population alone means that the loss of habitat for other organisms is perhaps the most pressing problem (Laurance et al., 2014), although what habitat remains is under tremendous pressure from fragmentation and degradation (Haddad et al., 2015). Furthermore, biodiversity is being "homogenized," or more colloquially put, being put in a blender, with rare species being exterminated, some due to overharvesting, whereas other invasive species are spreading throughout the world (Olden et al., 2004).

Ironically, as we lose biodiversity we are only just beginning to understand it. For community ecologists, this means that only now we are getting a better conception of how species interact, what forces underlie a diverse community, and what makes one community more resilient to disturbance than another. And as we come to understand the complexity of natural systems, it becomes all the more clear that the loss of biodiversity is not just the progressive loss of one species after another, but also the loss of the interactions between these species (Tylianakis et al., 2009). Indeed, these interactions between species might be more sensitive to human disturbance than the participating species themselves (Valiente-Banuet et al., 2015), and there is the possibility that linkages between species could lead to coextinction (Dunn et al., 2009). Of course, species are also not just affecting each other but also affecting the abiotic environment. The overall outcome of these interactions for humans is that communities of animals produce "ecosystem services" that affect human welfare in several ways, from the water we drink, to pest

Mixed-Species Groups of Animals. http://dx.doi.org/10.1016/B978-0-12-805355-3.00008-7
Copyright © 2017 Elsevier Inc. All rights reserved.
147

management, to the recycling through ecosystems of detritus and various elements and toxins Şekercioğlu (2006). All these reasons call for conservation of species interactions with each other and with the environment, not just the species themselves.

Species interactions are clearly fundamental to mixed-species associations (MSAs), and so in this chapter we review what we know about how MSAs respond to human disturbance and how they might best be managed and conserved. We will concentrate on avian mixed-species groups (MSGs), the most studied taxa from a conservation perspective, which have been suggested to have high conservation value simply because they often compose a large percentage of the local avifauna and thus are an easily observable indication of the health of the bird community at any one place (Lee et al., 2005). Yet although much of what we know is specific to MSGs, we will also try to incorporate MSAs when applicable, because, as we argued in Chapter 2, many of the principles that underlie the organization of MSA phenomena also are common to MSGs. In particular, we have seen in Chapters 2, 3, and 7 how species that provide public information, make food accessible, or act as sources of protection can play important "nuclear" or "keystone" roles in MSAs, and in MSGs specifically. Such species can be considered efficient targets for conservation because by conserving them one also positively affects other species (Simberloff, 1998). Indeed, we will argue that species' effects on other species should be integrated into their conservation value.

8.2 MIXED-SPECIES GROUP RESPONSES TO ANTHROPOMORPHIC DISTURBANCE

Early descriptions of avian MSGs included those in human disturbed habitats (e.g., Ulfstrand, 1975; Croxall, 1976); however, the focus of these articles was on a description of the group system or its relationship to vegetation, and not directly on how disturbance affects MSGs. This situation began to change in the early 1990s, when two long-term studies in the Neotropics—the Biological Dynamics of Forest Fragmentation in Brazil, a large-scale experimental study of fragmentation (Bierregaard and Lovejoy, 1989; Stouffer and Bierregaard, 1995), and the work of Jean-Marc Thiollay (1992) in French Guiana—both reported that MSG members were among the most vulnerable (i.e., prone to extirpation from an area) species to human disturbance. It should be remembered that both of these studies were focused on Amazonian understory avian MSG systems, similar to that described by Munn and Terborgh (1979), so their results cannot be viewed as being completely independent of each other (but see Sigel et al., 2006, for similar, more recent results from Central America). Nevertheless, this work spurred much further interest in MSG response to human disturbance.

Since then this subfield of the avian MSG literature has grown rapidly, so that currently there are about 20 articles to our knowledge (Fig. 1.2; Table 8.1).

TABLE 8.1 How Avian Mixes-Species Group (MSG) Characteristics Respond to Disturbance

Gradient	Reference	SR	NI	FP	FE	Other Response/Notes
Fragmentation						
	Poulsen (1994)	↓	↓	–	–	
	Maldonado-Coelho and Marini (2000)	↓	↓	–	–	Decrease in very small fragment in one season
	Tellería et al. (2001)	↓	↓	–	–	Scanning behavior increases
	Fernández-Juricic (2000, 2002)	0	↓	↓	–	
	Maldonado-Coelho and Marini (2004)	↓	↓	–	–	Declines, by only one MSG type, include reduced stability
	Van Houtan et al. (2006)	–	–	–	–	Over time, MSG species extirpated
	Sridhar and Shankar (2008)	0	0	–	→	
	Brandt et al. (2009)	–	–	–	–	Analysis centers on changing composition
	Mokross et al. (2014)	↑	–	–	→	Number of associations decreases in fragments
	Cordeiro et al. (2015)	↓	↓	–	–	Composition more variable in fragments
Land-Use Intensity						
	Pomara et al. (2003, 2007)	–	–	↓	–	Declines for two of four species
	Lee et al. (2005)	↓	–	–	–	

Continued

Table 8.1 How Avian Mixes-Species Group (MSG) Characteristics Respond to Disturbance—cont'd

Gradient	Reference	SR	NI	FP	FE	Other Response/Notes
	Matthysen et al. (2008)	↓	–	–	–	
	Sidhu et al. (2010), Goodale et al. (2014), and Mammides et al. (2015)	↓	↓	↓	↓	
	Knowlton and Graham (2011)	↓	↓	↓	–	Declines in one of two habitats
	Zhang et al. (2013)	↓	↓	↓	0	
	McDermott and Rodewald (2014)	↓	↓	–	–	
	Colorado and Rodewald (2015)	↓	↓	–	–	
Other						
(Roads)	Develey and Stouffer (2001)	–	–	–	0	Open road forms territory boundaries
(Edge vs. Interior)	Péron and Crochet (2009)	↑	–	–	–	

Only studies that directly investigated MSGs, rather than the bird community as a whole, are included. "0" a nonsignificant effect; "–" was not studied; SR, species richness; NI, number of individuals; FP, propensity to be in MSG; FE, MSG encounter rate or density. Up or down arrows represent significant effects.

Articles on similar issues for other taxa are as yet rare but undoubtedly will soon appear. For example, Tisovec et al. (2014) have recently explored how primate MSGs respond to agroforestry. For birds, the literature so far has described two related but distinct gradients: land-use intensity (changes over time or between land-use types) and habitat fragment size. A few articles have looked at slightly different aspects of change, including how MSGs respond to roads (Develey and Stouffer, 2001) or changes in MSG composition at the edge between natural and human disturbed habitat (Péron and Crochet, 2009). Noise pollution can even degrade communication between species, not allowing species to use other species' alarm calls (Grade and Sieving, 2016).

The findings from this work indicate that human-induced disturbances to avian MSG systems are largely negative (Table 8.1), although this is difficult to quantify due to the wide variety of such disturbances and the different kinds of responses measured. Most aspects of MSGs tend to be impoverished in more disturbed conditions, including species richness, number of individuals, group density, or grouping propensity, with the exact aspect that shows change differing from study to study. One study has even looked at how the frequency of association, as measured by social network statistics, decays over a fragmentation gradient (Mokross et al., 2014; Fig. 8.1). In some highly disturbed land-use types such as treeless agriculture (e.g., tea plantations), MSGs can completely disappear (Sidhu et al., 2010). Similarly, there appears to be a threshold of fragment size, somewhere between 1 and 10 ha depending on the area, in which groups can no longer persist in fragments (Stouffer and Bierregaard, 1995; Maldonado-Coelho and Marini, 2000).

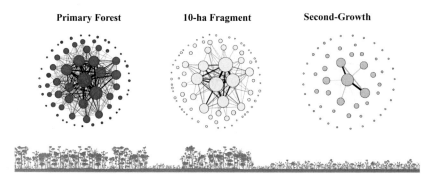

Primary Forest   10-ha Fragment   Second-Growth

FIGURE 8.1   **The social networks of birds in avian mixed-species groups (MSGs) decay over a disturbance gradient.** Avian MSGs were studied as part of the Biological Dynamics of Forest Fragments Project, in Amazonian Brazil, in three habitats: primary forest (left), 10 ha fragments (middle), and secondary forest (right). *Circles* represent species and lines are proportional to the interactions between species; the vegetational structure of the different environments is diagrammed at the bottom. *Figure used from Mokross, K., Ryder, T.B., Côrtes, M.C., Wolfe, J.D., Stouffer, P.C., 2014. Decay of interspecific avian flock networks along a disturbance gradient in Amazonia. Proceedings of the Royal Society of London B: Biological Sciences 281, 20132599, with permission.*

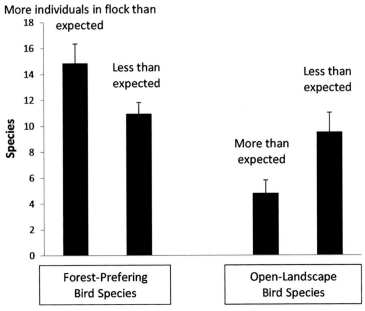

FIGURE 8.2   Avian mixed-species groups in disturbed habitats contain more forest interior species than expected by chance. The figure compares what bird species participate in groups outside of protected forest in Sri Lanka to what would be expected from random simulations of group composition based purely on the abundances of the species. For the category of forest-preferring species, the number of species that had more individuals in actual groups than simulated groups was significantly greater than the number of species that had fewer individuals. The opposite pattern was true for open-landscape species. *Adapted from Mammides, C., Chen, J., Goodale, U.M., Kotagama, S.W., Sidhu, S., Goodale, E., 2015. Does mixed-species flocking influence how birds respond to land-use intensity? Proceedings of the Royal Society of London B: Biological Sciences 282, 20151118.*

A subtle, but critical aspect of this issue is that even though MSG degrade in disturbed areas they still can be important to bird communities in such areas. For example, Mammides et al. (2015) showed that, although groups are less numerous outside of protected areas, forest-preferring birds that persist in such areas are more likely to be found in MSGs than would be simply expected from their abundance (Fig. 8.2). The study by Tisovec et al. (2014) showed that tamarins were more often in MSGs in more disturbed areas. In some very disturbed areas such as tea plantations, grassland MSG systems may replace the forest systems that were presumably there before the agriculture, providing important ecosystem services such as insect control (Sinu, 2011). Hence, although MSGs might not be abundant in disturbed areas, it is still useful to study how they can persist in or move through such lands.

8.3 MECHANISMS OF THIS RESPONSE

What are the exact factors that explain this general pattern of MSG degradation in disturbed environments, at least for birds? First, we should remember that avian MSGs are most dominant in forested environments. As forest structure degrades, it is probable that the predation risk regime typical of forests also changes and is perhaps replaced by a risk environment more typical of grassland habitats, which are expected to have looser and less diverse MSGs (although ones that may have a high number of individuals, Terborgh, 1990). A simpler structural environment may thus decrease the propensity of some bird species to join groups. Such an effect could interact with other outcomes of habitat fragmentation and degradation in complex ways. For example, fragmentation may reduce the number of insectivorous species due to declining food availability, loss of preferred microhabitats, decreased dispersal, and increased nest predation (Powell et al., 2015), and if species are exterminated by these mechanisms, they will not be available to join groups. Potentially there might be a threshold number of MSG participants, and if there are fewer species than this threshold, MSGs may not be able to form (Harrison Jones, personal communication).

An additional stress of MSGs in disturbed fragments could be that participants in groups require more space than other species and would thus drop out of groups in smaller fragments. This has been suggested for the Amazonian system in which several "core" group obligate species, never found outside of MSGs, mutually defend a territory (Van Houtan et al., 2006). Beyond these core species in this system, there are also some species that have larger territories than the group and switch between groups that reside in their territories (Munn and Terborgh, 1979; Jullien and Thiollay, 1998); such species would clearly be sensitive to area effects. Yet it is unclear whether space requirements for group participants are any different than other species in regions where MSGs do not appear to hold joint territories but rather move like a wave through the forest with individuals joining in and then dropping out as the group enters and exits their territories (McClure, 1967).

Another class of responses of MSGs to disturbance might be classified as "cascades," in which nuclear or leading species may influence following species. For example, it has been repeatedly suggested that when nuclear species become rare or absent, groups then dissolve (Dolby and Grubb, 1999; Maldonado-Coelho and Marini, 2004; Sridhar and Sankar, 2008; Zhang et al., 2013). Mammides et al. (2015) demonstrated the potential for such a cascade when they showed that two leading species differed in their sensitivity to land-use intensity and other species differed in which of the two leaders they preferred, with followers preferring leaders of similar body size. We say "the potential for such a cascade" because the impact of this effect is hard to disentangle from other factors. For example, Mammides et al. (2015) note that one large vulnerable species, the red-faced malkoha, was rarely found outside of groups led by

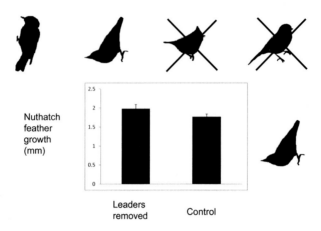

FIGURE 8.3 **An experimental study of the effect of group leaders on following species.** These North American groups usually consist of tufted titmouses and chickadees, which lead the flocks, and downy woodpeckers and white-breasted nuthatches, which follow; in the experiment the titmouses and chickadees were removed (top panel). Nuthatches in treatment groups saw lower feather growth than in control groups where the leaders were still present (bottom panel), suggesting a negative impact of the loss of leaders. *Adapted from Dolby, A.S., Grubb Jr., T.C., 1998. Benefits to satellite members in mixed-species foraging groups: an experimental analysis. Animal Behaviour 56, 501–509.*

the large leading species, the orange-billed babbler, and also that malkohas were infrequently found outside of forests. But whether this species' exclusivity to forests is actually caused by its dependence on babbler-led groups, which are in general found within protected forests, is unclear; other variables particular to this species may also play a large role in its habitat selection.

Cascades between leaders and followers in MSGs could either be negative or positive in terms of how they affect the sensitivity of followers to habitat disturbance, depending on the characteristics of the nuclear species. A nuclear species that is tolerant of human disturbance could make its followers more tolerant as well. The exact effect of leading species on followers is also shaped by how dependent the followers are on the leaders. There is unfortunately little data on this subject. In one elegant study, Dolby and Grubb (1998) showed experimentally that the removal of leading species resulted in poorer condition of individuals of a following species, although the effect size was relatively small (Fig. 8.3). This lack of data on the dependence of followers on leaders makes it unclear whether the targeting of nuclear species would be a truly effective conservation approach, a subject we return to later (Section 8.6).

One final mechanism by which MSG systems could influence their participants' response to human disturbance is that MSG participation may lessen predation risk and therefore allow animals to continue to use habitat they might otherwise avoid. A few experimental studies have shown that following species may be more likely to take risks and come from a forest edge into the open if leading species are present. One study was linked to the same experimental

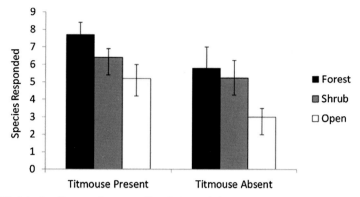

FIGURE 8.4 **Leading species may affect followers' risky movements into disturbed habitats.** Birds were more likely to inspect an owl model if a group leading species, the tufted titmouse, was present. Inspection behavior was also influenced by the habitat in which the model was placed. *Adapted from Sieving, K.E., Contreras, T.A., Maute, K.L., 2004. Heterospecific facilitation of forest boundary crossing by mobbing understory birds in north-central Florida. Auk 121, 738–751.*

"knockout" of MSG leaders described above; when the leader was present, a follower species used a feeding table in the open more willingly (Dolby and Grubb, 2000). In another study, bird species were more apt to come toward an owl model when the MSG leader was present (Sieving et al., 2004; Fig. 8.4). In addition, some observational studies have shown that birds may transition more between forest and other habitats when in MSGs (Tubelis et al., 2006; Péron and Crochet, 2009). These forces may operate at a small scale, determining movements of animals along an edge. In contrast, the idea that participants in MSGs may avoid treeless areas may explain MSG absence at a larger scale in highly disturbed continuous blocks of habitat.

8.4 GAPS IN KNOWLEDGE

We clearly know enough now about how avian MSGs respond to disturbance to take some immediate actions to protect them (Sections 8.5–8.7). Yet there are still gaps in our knowledge about this subject that should be filled to make the strongest case for community conservation. A major gap, in our view, is that there is not much evidence for how species' distributions or the fitness of individuals of different species are affected by MSGs. Obviously, if participation in an MSG has very little impact on fitness or on habitat use, there would be little incentive to direct conservation effort at the level of an MSG. Unfortunately, even for the better-known systems, there are few studies that have actually tried to determine the fitness consequences associated with the formation of MSGs. Only one study (Jullien and Clobert, 2000) has shown that survival is higher for species that are obligate group members (those species never seen outside groups in the Neotropics) but not for species that join groups occasionally, which suggests that the longevity benefits might differ according to the degree of participation in MSGs. There have been, however, several studies that looked at foraging rates inside and outside of groups (e.g., Hino, 1998; Pomara et al., 2003; Satischandra et al., 2007, and reviewed by Sridhar et al., 2009), and we expect that increased foraging efficiency, showed especially by flycatching species that follow flocks in these studies, should translate into positive fitness effects.

The influence of a leading species on other species is even more difficult to isolate. We have already discussed the studies of Dolby and Grubb (1998, 2000) in which leading species were removed from avian MSGs and the effects on the remaining species were measured. We believe this approach is particularly promising in giving strong data about species interdependencies. However, it is difficult to implement: one needs to have widely isolated forest patches in which birds can be removed without other birds immediately replacing them. Furthermore, there are some difficulties with interpretation. For example, if one removes a gregarious leader, are the remaining species affected by the loss of that species or by the groups being much smaller? Another option is to use species association statistics from observations of cooccurrence, as well as null models and network statistics, to assess and rank species' importance to MSGs (Sections 7.2.2.3 and 7.2.2.4). These methods can then be used to determine which species most strongly affect other species' participation in MSGs and also whether the roles of such "keystone" species in groups are consistent over a gradient of human disturbance.

MSGs' requirements for space and landscape features also need further study. Although there have been excellent studies in the Neotropics in which birds were individually identified (Munn and Terborgh, 1979; Gradwohl and Greenberg, 1980; Jullien and Thiollay, 1998), such "color-banded" studies have been very rarely conducted in the Old World (but see Farine and Milburn, 2013; Farine et al., 2015a). As yet, outside of the Neotropics, there is no evidence for the kind of mutually defended group territories that have been described in the color-banded studies cited above. If mutual group territories occur, estimates of their size are essential for designing conservation plans for avian MSGs and specifically to estimate minimum fragment sizes to protect. Even if such mutual territories are not found elsewhere, studies that estimate the home ranges of the leading species would be valuable to prioritize fragments for conservation.

Another recent paradigm of bird spatial ecology is the "translocation" experiment in which birds are removed from their territory and then tracked as they return back to their territory to determine their habitat preferences when moving through human disturbed areas. For example, Gillies and St. Clair (2008) compared two species of birds, a forest specialist and a forest generalist, in what habitats they moved through when returning to their territory. They found that the specialist species preferred to stay in forest or riparian corridors (50–150 m wide) and did not use fencerows of trees (15–30 m wide); in contrast,

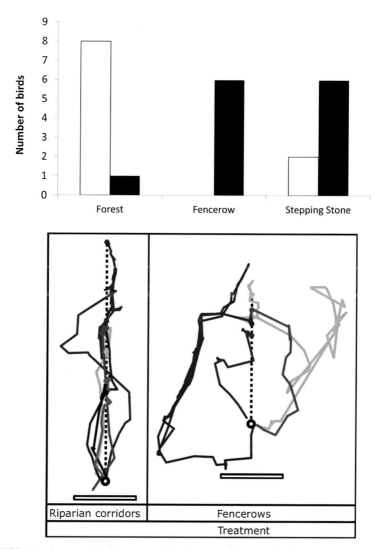

FIGURE 8.5 **An example of how a translocation experiment can indicate how animals move through disturbed habitats.** Two bird species, the barred antshrike (white), a forest specialist, and the rufous-naped wren (black), a forest generalist, used different habitats when they were experimentally translocated away from their territory and then tracked as they returned (top panel). Antshrikes would follow riparian corridors but not fencerows, as shown by the routes birds took in the bottom panel. *Used by permission from Gillies and St. Clair, copyright (2008) National Academy of Sciences, U.S.A.*

the generalist used fencerows more frequently (Fig. 8.5). This sort of information can tell us which parts of the habitat are important when animals move and how species may respond to habitat degradation. It would seem useful to try applying this experimental approach for MSG leaders to see what kinds of habitats or landscape features allow them to persist in or move through disturbed areas. Although such translocation methods have been most used so far in birds, they would also seem extendible to most taxa.

8.5 CONSERVATION ACTIONS: PROTECT LOCATIONS

We used the last chapter to promote further research, something that researchers are inclined to do, perhaps endlessly. But what should we do immediately to take conservation action for MSAs or MSGs? Let us first discuss location-based approaches. In Chapter 3, we mentioned several types of MSAs that are found in particular places. Generally, these are aggregations at the location of a resource, such as watering holes or mineral licks. This sort of landscape feature has been termed a "keystone structure" (Tews et al., 2004) and is obviously a target for conservation. At a smaller scale, this idea mirrors the idea of "biological hotspots," wherein whole geographic regions are identified as both biodiverse and threatened (Myers et al., 2000) and hence vital for conservation. Indeed, descriptions of such localized resources often emphasize that they should be prioritized for protection (e.g., Bravo et al., 2008). Several other kinds of MSAs discussed in Chapter 2, such as cleaning fish stations, mixed-species colonies or roosts, or constructions of animals such as prairie dog colonies, may have long histories of activity in one place and can also be mapped and protected. In these cases, this location-based approach is connected to a species-based approach, described below, investigating in what places certain species (e.g., prairie dogs) are particularly abundant.

A good example of location-based conservation strategies that focus on MSAs comes from fish. Due to overharvesting, fisheries are among the most affected groups from human activities, with current communities of fish and other edible coastal animals barely resembling what was present before exploitation (Jackson et al., 2001). Many species form fish spawning aggregations ("FSAs") and recent research has shown these sites, which are found in particularly productive areas, can be used by multiple species (e.g., Kobara et al., 2013). Hence it makes sense that FSAs should be a "focal point for marine conservation and fisheries management on a global scale" (Erisman et al., 2015).

8.6 CONSERVATION ACTIONS: PROTECT SPECIES

Currently most list or rankings of species (e.g., the IUCN Red List; http://www.iucnredlist.org/) that should be prioritized in conservation are based on those species' population sizes and distributions. However, protection on this basis alone may not be cost-effective. Species with small populations might require very expensive interventions with a low chance of success, and the success of a conservation project should be incorporated into the decision of whether to fund it, given the limited amount of money available for conservation (Joseph et al., 2009). Nevertheless, these rankings are currently widely used to plan reserve design, to constrain development, and to measure the state of the environment (Possingham et al., 2002).

Including species' effects on each other and the environment into the measuring and ranking of species' conservation value might increase the efficiency of conservation. Simberloff (1998) was among the first to call for this approach in his article "Flagships, umbrella, and keystones: is single-species management passé in the landscape era?" A wide range of species that play important roles in MSAs could be considered keystone species: initiators of aggregations, protective species, driving species, and leading species of moving groups. Indeed, there have already been calls for conserving protective species near which birds nest (Haemig, 2001) or conserving keystone species in fish nesting associations (Peoples and Frimpong, 2013). Special attention should be placed on such keystone species that themselves respond poorly to human disturbance, because their response could reverberate onto the other species they influence.

An example of how species interactions could be integrated into conservation plans is a proposed global ranking of MSG leaders by Fasheng Zou and colleagues working on an ongoing project. Using the large literature on avian forest MSGs, they have made a list of all species that have been described as "nuclear species" or leaders of this kind of MSG. They then ranked species using three scores: (1) the average number of other species that were described as "attendant" species or followers in MSGs these species led, (2) the consistency of leadership the species showed within their range (did they always lead MSGs?), and (3) the size of the species' ranges. The researchers hope that this list could have a role to play in setting species priorities for avian conservation.

8.7 CONSERVATION ACTIONS: RESTORATION OF DISTURBED AREAS

Conservation actions need not be confined to protected environments, especially because groups might be particularly important in disturbed areas. In Section 8.4 we mentioned that we need studies of what landscape features allow MSGs to persist and move in disturbed areas. With this information, we can then introduce these features back into a degraded area. For example, a fencerow of trees is an example of a landscape feature that is relatively low cost to build (in contrast, a riparian corridor might be even more important for conservation, but more expensive). Even if a fencerow was only used by some species (Gillies and St. Clair, 2008), if those species were important to MSGs it could have a large impact, connecting small fragments and perhaps putting them cumulatively over a threshold of size that would allow an MSG to stay in the area. And ultimately if we preserve and connect enough habitats, a keystone or leader species could be reintroduced into an area it had been exterminated from. These ideas are easy to write about but are undoubtedly much more difficult to implement on the ground, with a good partnership between ecologists and conservation practitioners necessary.

8.8 CONCLUSIONS

Currently, however, there often appears a wide gap between academic conservation biologists and conservation practitioners (Knight et al., 2008; Laurance et al., 2012). Academic conservation biologists are judged by their publication record, and the high-impact journals they publish in seek data sets of "global importance" (Mammides et al., 2016). To be a conservation practitioner, however, it is important to really understand the local environment, both the ecological one and the human socioeconomic one. For example, in Section 8.7, we mentioned very briefly reforestation and reintroduction. Effective work in this realm is all about the specifics of the case. For example, in the construction of a corridor, how wide should that corridor be? Using what plant species? The answers to these concerns will be found out through assessment on what is important for the local fauna and what is economically viable for the local people to accept the restoration project as a part of their sustainable development. Yet the answers to all these crucial questions may be of little interest to academic journals. At the same time, issues that seem important to academics because they are found throughout the world might pale in importance to these local factors when actually planning conservation on the ground.

Hence we think there is a danger that we, as ecologists, let our own academic interest in MSGs drive us to overemphasize this one issue in the complex world of conservation practice. Should the fact that animals interact in MSGs be considered in conservation? We argue confidentially yes in this chapter but admit that it is hard to put a weight on this one characteristic. To make the best case of MSA/MSG community conservation, we think it is especially important to have more data on the dependence of participating species on the MSG systems and the nuclear species in those systems. We also hope for a fruitful dialog with practitioners and wish they will be able to communicate both their successes and failures through forums for publishing such results such as Conservation Evidence (http://www.conservationevidence.com/).

As a final word on the connection between MSA/MSGs and conservation, let us return to the point made in the preface of how these systems are very exciting for people to experience and thus have a role in education about nature and why it is important to conserve. We have emphasized in this chapter nuclear species as "keystones" due to their effects on other species and MSGs as "indicators" of environmental health. Yet because of their ability to inspire and connect people with nature, MSGs may also be appropriate "flagships" for conservation (Simberloff, 1998; Caro and Girling, 2010), the attention-grabbing species whose conservation can lead to the protection of many other species in the places where they live.

Chapter 9

Conclusions

9.1 WHAT HAVE WE LEARNED?

One of our motivations in writing this book is that the vast literature that has accumulated over time about mixed-species groups (MSGs) lacks a simple yet detailed introduction across taxa to the subject. We hope our review of the existing literature will provide such a platform on which to base future research. Below, we provide some generalizations that have emerged when writing the book and highlight some new directions that could prove productive. We organize the chapter in a similar manner to the book, covering first the empirical description of MSGs, then the theories of adaptive benefits to MSGs, then the behaviors that shape species interactions in MSGs, and finally conservation.

9.2 EMPIRICAL RESEARCH ON MIXED-SPECIES GROUPS

The experience of reviewing the existing literature on what species participate in MSGs (see in particular Chapter 3) has been eye-opening in that it really shows birds as the one vertebrate taxon broadly involved in MSGs, with many families involved under different ecological settings (from aquatic to terrestrial) and a high percentage of the species involved at any one place. This is not just limited to MSGs but also applies to mixed-species associations (MSAs) more generally: waterbirds are often found in moving aggregations; birds are also prime examples of nesting near protective species and of mixed-species colonies. Given this domination of mixed-species phenomena by birds, we may ask then what is it about birds that makes mixed-species interactions so frequent?

One answer to this comparative domination by birds focuses on aspects of other taxa that are not as compatible with an MSG "lifestyle." Of three major vertebrate groups considered in this book, mammals are simply the least evenly social: that is, sociality is found only in certain lineages, although it is well developed in those lineages. The species-rich taxa of mammals, such as rodents and bats, are not ones as yet highlighted in the MSG literature, although we note in passing that many texts about bats give the impression that mixed-species colonies are common, and we hope that more empirical work will focus on mapping the relative positions of species in caves. For fish, it seems that the emergent properties of MSGs in fish may be more complicated than those for birds. The fish MSG may exist as a more cohesive, tightly packed unit than the bird MSG, in which individual bird movements while

Mixed-Species Groups of Animals. http://dx.doi.org/10.1016/B978-0-12-805355-3.00009-9
Copyright © 2017 Elsevier Inc. All rights reserved.
161

foraging might be quite uncoordinated with others. For example, although fish may swim together without foraging, this is rare for birds. One example recently well described involves groups of two species of crows in England that fly together to their roost (Jolles et al., 2013). Indeed, in this situation, the individuals show preferences to associate and cluster with conspecifics within the overall MSG, similar to the behavior of fish. The cohesiveness of fish MSGs may then enforce stricter rules on what individuals are involved and how similar they need to be to each other. It is not clear to us whether such rules are enforced by the predators (e.g., the oddity effect) or whether they might have to do with the kinetics of fish movement and how energetically efficient coordinated movement can be (Krause et al., 1996).

By contrast to fish and mammals, social foraging is well developed across many avian families (Beauchamp, 2002). At least when foraging, many avian species exploit large food patches that can easily be shared. Certainly a first step in the evolution of MSAs or MSGs is the ability to form groups or tolerate neighbors. A close second step would be the ability to exploit similar resources. Those two requirements are met in many avian species. Our review of the costs and benefits associated with MSG formation also shows that individuals in such groups can achieve a higher foraging efficiency and experience a reduction in predation risk. With greater opportunity to forage together and share the benefits that ensue from the formation of groups, birds are thus in a position to form MSAs and MSGs more effectively. In addition, their ability to move great distances rapidly also probably facilitates the formation of groups in birds.

One of the major gaps in knowledge we have encountered is the lack of integration of invertebrates in the mixed-species literature. As described before (Section 3.2), part of the reason invertebrates are given cursory mention in this book is due to the way we delineate its scope, avoiding symbioses, and also to differences in scale, for example, those that alter the way we see the interaction of pathogens, parasites, and crop pests in hosts. Nevertheless, pockets of strength in the literature, for example in small crustaceans and whirligig beetles, are intriguing and suggest that some taxa may be overlooked in the context of MSGs and MSAs. In general, we hope that future tests of the adaptive hypotheses of MSGs will use a single currency, such as risk of predation, to analyze multiple taxa. In addition, our knowledge has proceeded to the level where the lack of MSGs in a particular taxon or place may not simply reflect ignorance or neglect but may point the way to why MSGs form in other conditions (e.g., Willis, 1973a).

The advancing technology available for bioinformatics also provides new opportunities for empirical studies. For example, citizen science initiatives such as eBird (Wood et al., 2011) allow amateur scientists to collect information with some oversight from experts. In our viewpoint, information about interactions between species in MSGs could be fairly easily integrated into such platforms. Indeed, Eben Goodale and Guy Beauchamp were among 35 scientists who approached eBird in 2014 with the idea of incorporating MSG participation into the information they gather. We argued that given some reasonable standardization about how MSGs are defined, information on their composition could be integrated into the "details" section that eBird has for each bird observation. Problems involving computer coding and funding have yet, however, to allow progress in these directions. Although the number of descriptive studies as a proportion of the total research activity on MSGs has declined (Section 1.2), and likely will continue to, descriptive information from new localities, especially gathered at the volume allowed by citizen science, could revolutionize the field by changing the grain—the pixel size—of the information we have about MSGs (Fig. 9.1). Furthermore, repeated observations at localities previously surveyed (e.g., Martínez and Gomez, 2013) could reveal how MSGs are responding to human-induced disturbances and climate change.

9.3 THEORETICAL REVIEW OF BENEFITS AND COSTS

Significant gaps remain in our understanding of the costs and benefits to individuals of participation in MSGs (Chapters 4 and 5). However, in general, there is no fundamental impediment to closing these gaps, and empirical work can take advantage of the growing availability of miniature sensors carried on-board animals (as well as drone technologies) to provide very detailed behavioral data (Fig. 9.1). As an example, and as discussed in Section 4.2.7, we have a very poor understanding of the costs and benefits of MSG formation as a means of defending an exclusive territory. Within such a group-held territory, it is unclear whether individuals perform better during foraging by remaining together or by separating and all foraging separately. Furthermore, such foraging consequences need to be integrated with costs and benefits in terms of continued

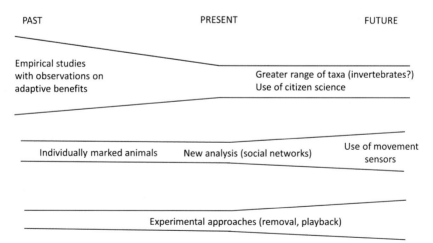

FIGURE 9.1   The methodology used in the study of mixed-species groups: past, present, and future. We foresee a proportional reduction in the amount of descriptive papers of unmarked individuals, with more experimental approaches and movement studies of marked individuals.

defense of the territory. By separating during foraging the territory holders will be spread across the territory more and so will be able to detect encroachment by others more quickly. However, by being initially spread out, the collective ability of the territory holders to repel any detected encroachers may be reduced until such time as they reform into a coherent group. Given that many primates and bird species are often territorial but form MSGs these issues should be amenable to empirical study.

Predation in vertebrates is challenging to study. Firstly, it is logistically and ethically challenging to explore empirically in the laboratory. Secondly, it is onerous to collect data on naturally occurring predation on vertebrates, as instances are difficult to predict in space and time, and easily influenced by the presence of human observers. This is highly problematic for our understanding of MSGs because most reported instances of MSGs involve vertebrates and predation pressure appears to be a strong driver of grouping behavior across the natural world. We see this problem laid out in Chapter 5. For example, attack abatement (Section 5.1.1), nonrandom prey choice within groups by predators (Section 5.1.4), and social learning about predators (Section 5.1.6) are all conspicuously understudied in the context on MSGs—despite all these been strongly implicated in the evolution of single-species groups. However, the increasing ability to track free living vertebrates using data loggers and on-board cameras is already revolutionizing our ability to study animal behavior and will surely help to close these gaps in our current knowledge.

Chapter 5 also highlights a number of case-studies of MSAs where costs and benefits accruing from these associations have been quantified. However, even in cases where the systems have become staples of textbooks and TV documentaries, we find that crucial information is missing for us to construct a full picture of the selection pressures acting on all parties in the interaction. The most striking case of this comes in the association between some seabirds, dolphins, and tuna engaged in hunting groups of fish. This system has been well known to human fishermen for decades, who have used seabirds as a reliable cue for finding tuna to net. However, we have very little data on the consequences for any of the interacting species in these aggregations. Happily, the large body sizes of all the species involved means that carrying current generation tracking and data logging equipment would be straightforward. However, the technological challenge to overcome would be how to affix such tags in the first place, although this challenge may be reduced if the dolphins and seabirds especially are attracted to human vessels.

We discuss MSAs between plants in Section 4.3 and establish that nonrandom associations between species are not uncommon (especially in harsh environments) and that mechanisms underlying benefits to association can be identified. However, evolutionary aspects of this phenomenon are generally ignored, perhaps because it is assumed that plants' very limited influence on the microhabitat in which their seeds settle precludes adaptation. However, we believe that selection for germination to be sensitive to the proximity of other species and for established plants to influence the germination of seeds of other species in their immediate vicinity merits further investigation.

In summary, one bitter–sweet issue that has emerged from writing this book is that we predict a technology-driven step change in data collection that should improve our understanding of MSAs considerably. We hope that this book will help with the theoretical framing of such studies so that data are collected and used to the best effect, but we expect this book to become dated as new data accumulate. We are looking forward to write a new edition to chart progress.

9.4 SOCIAL INTERACTIONS WITHIN MIXED-SPECIES GROUPS

Research on MSGs initially focused on documenting the bewildering diversity of species living in such groups. Evolutionary questions about the costs and benefits of living in MSGs also arose quite early. These questions are important to understand the selection pressures currently faced by species joining MSGs and ultimately to understand how such groups can evolve. Such questions have been covered in Chapters 4 and 5.

Observations on deceptive communication and roles in MSGs (Chapters 6 and 7) have also led researchers to wonder whether living in such groups can be an evolutionary force that shapes not only anatomical features but also behavioral attributes (Moynihan, 1960; Harrison and Whitehouse, 2011). The first important issue that remains to be evaluated is whether particular traits expressed by a species in an MSG, such as leadership, color patterns or vocalizations, are a response to the particular challenges that arise from living in an MSG or simply represent traits that have evolved in other contexts and that are coopted in the MSG context.

This issue can be tackled in several ways. In particular, we note that traits that are uniquely expressed in an MSG context, and not for instance in the single-species groups of the same species, are likely the consequence of unique selection pressures acting in an MSG. The use of deceptive communication by some species in MSGs appears a prime example of just such a trait because cheaters are almost always solitary outside the MSG. Comparative analyses could also be used fruitfully in future endeavors. For example, we describe in Chapter 7 that species that lead MSGs tend to occur in larger groups than other species that typically follow them. A large group size might simply be a response to selection pressures acting when the species forages in single-species groups. In this case, the role of leader is facilitated by the large group size, but a large group size may not be a trait that evolved in the context of an MSG. This issue could be addressed by matching leader species to sister species that are closely related but that do not associate with other species. If sister species are consistently found in smaller groups than matched leader species, we could argue that a large group size represents an adaptation to some of the unique challenges faced by species in an MSG context.

Recently, it has been suggested that by joining MSGs, species create a novel social environment that can shape not only their own behavior and morphology but also that of the other species that are present as well (Harrison and Whitehouse, 2011; Farine et al., 2015b). As such, living in an MSG could be seen as niche construction, the phenomenon by which species are not only simply adjusting to selection pressures from their environment but also shaping the environment and thus the selection regimes experienced by their descendants and other species that occur in the same environment (Odling-Smee et al., 2003).

We would like to stress that not all adaptations to MSG living, if such adaptations can be unambiguously identified as we discussed earlier, should be viewed as instances of niche construction. Consider the following simple example: Species A occupies the canopy in the forest and species B lives in the lower strata. Over evolutionary times, the two species form MSGs in the canopy when species B joins species A. Because there is more light in the canopy, species B gradually undergoes changes in coloration and perhaps even converges on the color pattern exhibited by species A. To us, these changes in appearance over time are not necessarily a case of niche construction but could simply be an evolutionary response to a niche change caused by the formation of MSGs. To invoke niche construction, we would have to show that heritable variation in species A is relevant to the changes in appearance in species B.

In contrast, now consider the case in which species B is now able to exploit the food discoveries of species A. If all individuals of species B scrounge their food instead of searching independently for the benefit of everyone, fewer resources will be uncovered and competition will intensify to the point where individuals of species A could actually evolve tactics to minimize thefts. Over evolutionary times, species B could evolve a level of food usurpation that provides enough dividends to compensate for any potential costs associated with countertactics adopted by species A. This could be viewed as an instance of niche construction because heritable variation in the scrounging behavior of species B can have an impact on the fitness of its own offspring (by allowing them to be more successful themselves in the future) and crucially on that of species A as well. In short, species B can modify some traits of species A over time because the success or failure of individuals of species B has an impact on the fitness of individuals of species A. In some sense, species A is now part of the extended phenotype of species B (Dawkins, 1982), and the group and its composition has influenced the evolution of the species inside it (Farine et al., 2015b). It is in this strict sense that we view adaptations to MSG living as niche construction or as an extended phenotype. We foresee that this distinction will become an issue as more cases of putative adaptations to MSG living are discussed in the literature.

9.5 CONSERVATION AND CONCLUSION

We hope that the recent flurry of research that has focused on the conservation implications of avian MSAs and MSGs will soon spread to other taxa. In particular, several types of primates have particularly close-knit groups, and MSGs are frequent in coral reef fish, which inhabit a highly threatened environment (Baker et al., 2008). Given the high percentage of avifauna involved in MSAs/MSGs in a variety of habitats, emphasizing MSGs as a community-wide conservation strategy will work most naturally for birds. In reviewing the literature for Chapter 8, we come to the conclusion that for forest avian MSGs in particular, there is truly the basis for such a community approach. Forest birds are known to be sensitive to disturbance overall (Gibson et al., 2011), but beyond that, species that participate in MSGs are exceptionally vulnerable (see Table 8.1). MSGs thus might be an example where species interaction networks may be even more sensitive to anthropogenic disturbance than the participating species themselves (Valiente-Banuet et al., 2015). We also have a good understanding of social interactions between species in general and what makes a species nuclear to MSGs (Chapters 6 and 7), so that we can target the system as a whole and not just the individual species.

In Section 8.4, we have already discussed in detail a number of research topics that need further investigation. The actual dependencies of species on each other, measured by how their fitness changes due to the interaction, is of major concern is understanding how much emphasis should be put on conserving the MSG system. We also need to have better understanding of how MSGs use space and select habitats, especially in degraded landscapes.

But ultimately, as we argue in Sections 8.5–8.7, we know enough already to take conservation action. We thus hope that any update of this text in years to come would showcase studies in which our knowledge of MSAs/MSGs were used to (1) assess the ecological health of an area, (2) to select what areas are appropriate to be conserved as protected areas or sites actively managed by the local community, and (3) to target specific keystone species whose protection can ensure that other species too will continue to thrive. If we do not start this process soon, MSGs will become rarer, less observable systems. The other future research projects discussed in this chapter would then be less achievable, and more importantly, less relevant to people, scientists and nonscientists alike, throughout the world.

References

Acevedo-Gutiérrez, A., DiBerardinis, A., Larkin, S., Larkin, K., Forestell, P., 2005. Social interactions between tucuxis and bottlenose dolphins in Gandoca-Manzanillo, Costa Rica. Latin American Journal of Aquatic Mammals 4, 49–54.

Agnew, W., Uresk, D.W., Hansen, R.M., 1986. Flora and fauna associated with prairie dog colonies and adjacent ungrazed mixed-grass prairie in western South Dakota. Journal of Range Management 39, 135–139.

Aivaz, A.N., Ruckstuhl, K.E., 2011. Costs of behavioral synchrony as a potential driver behind size-assorted grouping. Behavioral Ecology 22, 1353–1363.

Alatalo, R.V., 1981. Interspecific competition in tits *Parus* spp. and the Goldcrest *Regulus regulus*: foraging shifts in multispecific flocks. Oikos 37, 335–344.

Alevizon, W.S., 1976. Mixed schooling and its possible significance in a tropical western Atlantic parrotfish and surgeonfish. Copeia 1976, 796–798.

Allan, J., Pitcher, T., 1986. Species segregation during predator evasion in cyprinid fish shoals. Freshwater Biology 16, 653–659.

Allan, J.R., 1986. The influence of species composition on behaviour in mixed-species cyprinid shoals. Journal of Fish Biology 29, 97–106.

Allen, J.A., Anderson, K.P., 1984. Selection by passerine birds is anti–apostatic at high prey density. Biological Journal of the Linnean Society 23, 237–246.

Almany, G.R., Peacock, L.F., Syms, C., McCormick, M.I., Jones, G.P., 2007. Predators target rare prey in coral reef fish assemblages. Oecologia 152, 751–761.

Almany, G.R., Webster, M.S., 2004. Odd species out as predators reduce diversity of coral-reef fishes. Ecology 85, 2933–2937.

Alves, M.A.S., Cavalcanti, R.B., 1996. Sentinel behavior, seasonality, and the structure of bird flocks in a Brazilian savanna. Ornitologia Neotropical 7, 43–51.

Ancillotto, L., Allegrini, C., Serangeli, M.T., Jones, G., Russo, D., 2015. Sociality across species: spatial proximity of newborn bats promotes heterospecific social bonding. Behavioral Ecology 26, 293–299.

Anderson, M.G., 1974. American Coots feeding in association with Canvasbacks. Wilson Bulletin 86, 462–463.

Anderwald, P., Evans, P.G.H., Gygax, L., Hoelzel, A.R., 2011. Role of feeding strategies in seabird–minke whale associations. Marine Ecology Progress Series 424, 219–227.

Anguita, C., Simeone, A., 2015. Influence of seasonal food availability on the dynamics of seabird feeding flocks at a coastal upwelling area. PLOS ONE 10, e0131327.

Anguita, C., Simeone, A., 2016. The shifting roles of intrinsic traits in determining seasonal feeding flock composition in seabirds. Behavioral Ecology 27, 501–507.

Arbeláez-Cortés, E., Rodríguez-Correa, H.A., Restrepo-Chica, M., 2011. Mixed bird flocks: patterns of activity and species composition in a region of the central Andes of Colombia. Revista Mexicana de Biodiversidad 82, 639–651.

169

Ashmole, M.J., 1970. Feeding of western and semipalmated sandpipers in Peruvian winter quarters. Auk 87, 131–135.

Astaras, C., Krause, S., Mattner, L., Rehse, C., Waltert, M., 2011. Associations between the Drill (*Mandrillus leucophaeus*) and sympatric monkeys in Korup National Park, Cameroon. American Journal of Primatology 73, 127–134.

Au, D.W.K., 1991. Polyspecific nature of tuna schools: shark, dolphin, and seabird associates. Fishery Bulletin 89, 343–354.

Au, D.W.K., Pitman, R.L., 1986. Seabird interactions with dolphins and tuna in the eastern tropical Pacific. Condor 88, 304–317.

Austin, G.T., Smith, E.L., 1972. Winter foraging ecology of mixed insectivorous bird flocks in oak woodland in southern Arizona. Condor 74, 17–24.

Baigrie, B.D., Thompson, A.M., Flower, T.P., 2014. Interspecific signalling between mutual-ists: food-thieving drongos use a cooperative sentinel call to manipulate foraging partners. Proceedings of the Royal Society of London B: Biological Sciences 281, 41232.

Bailey, I., Myatt, J.P., Wilson, A.M., 2013. Group hunting within the carnivora: physiological, cognitive and environmental influences on strategy and cooperation. Behavioral Ecology and Sociobiology 67, 1–17.

Bailey, R.O., Batt, B.D.J., 1974. Hierarchy of waterfowl feeding with whistling swans. Auk 91, 488–493.

Baird, T.A., 1983. Influence of social and predatory stimuli on the air-breathing behavior of the African clawed frog, *Xenopus laevis*. Copeia 1983, 411–420.

Baird, T.A., 1993. A new heterospecific foraging association between the puddingwife wrasse, *Halichoeres radiatus*, and the bar jack, *Caranx ruber*: evaluation of the foraging consequences. Environmental Biology of Fishes 38, 393–397.

Bairos-Novak, K.R., Crook, K.A., Davoren, G.K., 2015. Relative importance of local enhancement as a search strategy for breeding seabirds: an experimental approach. Animal Behaviour 106, 71–78.

Baker, A.C., Glynn, P.W., Riegl, B., 2008. Climate change and coral reef bleaching: an ecological assessment of long-term impacts, recovery trends and future outlook. Estuarine Coastal and Shelf Science 80, 435–471.

Balda, R.P., Bateman, G.C., Foster, F.G., 1972. Flocking associates of the pinon jay. Wilson Bulletin 84, 60–76.

Ballance, L.T., Pitman, R.L., Reilly, S.B., 1997. Seabird community structure along a productivity gradient: importance of competition and energetic constraint. Ecology 78, 1502–1518.

Barbosa, P., Hines, J., Kaplan, I., Martinson, H., Szczepaniec, A., Szendrei, Z., 2009. Associational resistance and associational susceptibility: having right or wrong neighbors. Annual Review of Ecology, Evolution, and Systematics 40, 1–20.

Barnard, C.J., 1979. Predation and the evolution of social mimicry in birds. American Naturalist 113, 613–618.

Barnard, C.J., Sibly, R.M., 1981. Producers and scroungers: a general model and its application to captive flocks of house sparrows. Animal Behaviour 29, 543–550.

Barnard, C.J., Stephens, H., 1983. Costs and benefits of single and mixed species flocking in field-fares (*Turdus pilaris*) and redwings (*T. iliacus*). Behaviour 84, 91–123.

Barnard, C.J., Thompson, D.B.A., 1982. Time budgets, feeding efficiency and flock dynamics in mixed-species flocks of lapwings, golden plovers and gulls. Behaviour 80, 44–69.

Barnosky, A., Matzke, N., Tomiya, S., Wogan, G.O.U., Swartz, B., Quental, T.B., Marshall, C., McGuire, J.L., Lindsey, E.L., Maguire, K.C., Mersey, B., Ferrer, E.A., 2011. Has the Earth's sixth mass extinction already arrived? Nature 471, 51–57.

Bates, H.W., 1863. The Naturalist on the River Amazons. Murray Press, London.

Battley, P.F., Poot, M., Wiersma, P., Gordon, C., Ntiamoa-Baidu, Y., Piersma, T., 2003. Social foraging by waterbirds in shallow coastal lagoons in Ghana. Waterbirds 26, 26–34.

Bearzi, M., 2005. Dolphin sympatric ecology. Marine Biology Research 1, 165–175.

Beauchamp, G., 1999. The evolution of communal roosting in birds: origins and secondary losses. Behavioral Ecology 10, 675–687.

Beauchamp, G., 2002. Higher-level evolution of intraspecific flock-feeding in birds. Behavioral Ecology and Sociobiology 51, 480–487.

Beauchamp, G., 2008. A spatial model of producing and scrounging. Animal Behaviour 76, 1935–1942.

Beauchamp, G., 2014. Social Predation: How Group Living Benefits Predators and Prey. Academic Press, London.

Beauchamp, G., 2015. Animal Vigilance. Academic Press, London.

Beauchamp, G., Goodale, E., 2011. Plumage mimicry in avian mixed-species flocks: more or less than meets the eye? Auk 128, 487–496.

Beauchamp, G., Heeb, P., 2001. Social foraging and the evolution of white plumage. Evolutionary Ecology Research 3, 703–720.

Beauchamp, G., Ruxton, G.D., 2005. Harvesting resources in groups or alone: the case of renewing patches. Behavioral Ecology 16, 989–993.

Beauchamp, G., Ruxton, G.D., 2014. Frequency-dependent conspecific attraction to food patches. Biology Letters 10, 20140522.

Bednekoff, P.A., 2015. Sentinel behavior: a review and prospectus. Advances in the Study of Behavior 47, 115–145.

Beier, P., Tungbani, A.I., 2006. Nesting with the wasp *Ropalidia cincta* increases nest success of red-cheeked cordonbleu *Uraeginthus bengalus* in Ghana. Auk 123, 1022–1037.

Béland, P., 1977. Mimicry in orioles in south-eastern Queensland. Emu 77, 215–218.

Bell, H.L., 1983. A bird community of lowland rainforest in New Guinea. 5. Mixed-species feeding flocks. Emu 82, 256–275.

Bell, M.B.V., Radford, A.N., Rose, R., Wade, H.M., Ridley, A.R., 2009. The value of constant surveillance in a risky environment. Proceedings of the Royal Society of London B: Biological Sciences 276, 2997–3005.

Belt, T.W., 1874. The Naturalist in Nicaragua. Murray Press, London.

Bennett, J., Smithson, W.S., 2001. Feeding associations between snowy egrets and red-breasted mergansers. Waterbirds 24, 125–128.

Bernstein, I.S., 1967. Intertaxa interactions in a Malayan primate community. Folia Primatologica 7, 198–207.

Berthier, P., Excoffier, L., Ruedi, M., 2006. Recurrent replacement of mtDNA and cryptic hybridization between two sibling bat species *Myotis myotis* and *Myotis blythii*. Proceedings of the Royal Society of London B: Biological Sciences 273, 3101–3123.

Beyer, K., Gozlan, R.E., Copp, G.H., 2010. Social network properties within a fish assemblage invaded by non-native sunbleak *Leucaspius delineatus*. Ecological Modelling 221, 2118–2122.

Biani, N.B., Mueller, U.G., Wcislo, W.T., 2009. Cleaner mites: sanitary mutualism in the miniature ecosystem of neotropical bees nests. American Naturalist 173, 841–847.

Bierregaard, R.O., Lovejoy, T.E., 1989. Effects of forest fragmentation on Amazonian understory bird communities. Acta Amazônica 19, 215–241.

Bijlsma, R., 1984. On the breeding association between Woodpigeons *Columba palumbus* and Hobbies *Falco subbuteo*. Limosa 57, 133–139.

Binz, H., Foitzik, S., Staab, F., Menzel, F., 2014. The chemistry of competition: exploitation of heterospecific cues depends on the dominance rank in the community. Animal Behaviour 94, 45–53.

Bishop, A.L., Bishop, R.P., 2014. Resistance of wild African ungulates to foraging by red-billed oxpeckers (*Buphagus erythrorhynchus*): evidence that this behaviour modulates a potentially parasitic interaction. African Journal of Ecology 52, 103–110.

Bogliani, G., Sergio, F., Tavecchia, G., 1999. Woodpigeons nesting in association with hobby falcons: advantages and choice rules. Animal Behaviour 57, 125–131.

Boinski, S., 1989. Why don't *Saimiri oerstedii* and *Cebus cappucinus* form mixed-species groups? International Journal of Primatology 10, 103–114.

Botero, C.A., 2002. Is the white-flanked antwren (Formicariidae : *Myrmotherula axillaris*) a nuclear species in mixed-species flocks? A field experiment. Journal of Field Ornithology 73, 74–81.

Boulay, J., Deneubourg, J.L., Hedouin, V., Charabidze, D., 2016. Interspecific shared collective decision-making in two forensically important species. Proceedings of the Royal Society of London B: Biological Sciences 283, 20152676.

Bradbury, J.W., Vehrencamp, S.L., 2011. Principles of Animal Communication, second ed. Sinauer, Sunderland, MA.

Brady, S.G., 2003. Evolution of the army ant syndrome: the origin and long-term evolutionary stasis of a complex of behavioral and reproductive adaptations. Proceedings of the National Academy of Sciences of the United States of America 100, 6575–6579.

Brandt, C.S., Hasenack, H., Laps, R.R., Hartz, S.M., 2009. Composition of mixed-species bird flocks in forest fragments of southern Brazil. Zoologia 26, 488–498.

Bravo, A., Harms, K.E., Stevens, R.D., Emmons, L.H., 2008. Collpas: activity hotspots for frugivorous bats (Phyllostomidae) in the Peruvian Amazon. Biotropica 40, 203–210.

Brown, G.E., 2003. Learning about danger: chemical alarm cues and local risk assessment in prey fishes. Fish and Fisheries 4, 227–234.

Brown, G.E., Chivers, D.P., Smith, R.J.F., 1995. Fathead minnows avoid conspecific and heterospecific alarm pheromones in the faeces of northern pike. Journal of Fish Biology 47, 387–393.

Brumfield, R.T., Tello, J.G., Cheviron, Z.A., Carling, M.D., Crochet, N., Rosenberg, K.V., 2007. Phylogenetic conservatism and antiquity of a tropical specialization: army-ant-following in the typical antbirds (Thamnophilidae). Molecular Phylogenetics and Evolution 45, 1–13.

Brune, A., Friedrich, M., 2000. Microecology of the termite gut: structure and function on a microscale. Current Opinion in Microbiology 3, 263–269.

Bryan, R.D., Wunder, M.B., 2014. Western Burrowing Owls (*Athene cunicularia hypugaea*) eavesdrop on alarm calls of Black-Tailed Prairie Dogs (*Cynomys ludovicianus*). Ethology 120, 180–188.

Bshary, R., Grutter, A.S., Willener, A.S.T., Leimar, O., 2008. Pairs of cooperating cleaner fish provide better service quality than singletons. Nature 455, 964–966.

Bshary, R., Hohner, A., Ait-el-Djoudi, K., Fricke, H., 2006. Interspecific communicative and coordinated hunting between groupers and giant moray eels in the Red Sea. PLoS Biology 4, e431.

Bshary, R., Noë, R., 1997. Red colobus and Diana monkeys provide mutual protection against predators. Animal Behaviour 54, 1461–1474.

Buchanan-Smith, H.M., 1999. Tamarin polyspecific associations: forest utilization and stability of mixed-species groups. Primates 40, 233–247.

Burger, A.E., Hitchcock, C.L., Davoren, G.K., 2004. Spatial aggregations of seabirds and their prey on the continental shelf off SW Vancouver Island. Marine Ecology Progress Series 283, 279–292.

Burger, J., 1981. A model for the evolution of mixed-species colonies of Ciconiiformes. Quarterly Review of Biology 56, 143–167.

Burger, J., 1984. Grebes nesting in gull colonies: protective associations and early warning. American Naturalist 123, 327–337.

Burger, J., Gladstone, D., Hahn, D.C., Miller, L.M., 1977a. Intraspecific and interspecific interactions at a mixed species roost of Ciconiiformes in San Blas, Mexico. Biology of Behaviour 2, 309–327.

Burger, J., Gochfeld, M., 1994. Vigilance in African mammals – Differences among mothers, other females, and males. Behaviour 131, 153–169.

Burger, J., Howe, M.A., Hahn, D.C., Chase, J., 1977b. Effects of tide cycles on habitat selection and habitat partitioning by migrating shorebirds. Auk 94, 743–758.

Burtt, E.H., Gatz, A.J., 1982. Color convergence: is it only mimetic? American Naturalist 119, 738–740.

Buskirk, W.H., 1976. Social systems in a tropical forest avifauna. American Naturalist 110, 293–310.

Buskirk, W.H., Powell, G.V., Wittenberger, J.F., Buskirk, W.H., Powell, T.V., 1972. Interspecific bird flocks in tropical highland Panama. Auk 89, 612–624.

Buzzard, P.J., 2010. Polyspecific associations of *Cercopithecus campbelli* and *C. petaurista* with *C. diana*: what are the costs and benefits? Primates 51, 307–314.

Byrkjedal, I., 1987. Short-billed Dowitchers associate closely with Lesser Golden-Plovers. Wilson Bulletin 99, 494–495.

Byrkjedal, I., Eldøy, S., Grundetjern, S., Løyning, M.K., 1997. Feeding associations between Red-necked Grebes *Podiceps griseigena* and Velvet Scoters *Melanitta fusca* in winter. Ibis 139, 45–50.

Byrkjedal, I., Kålås, J.A., 1983. Plover's page turns into plover's parasite: a look at the Dunlin/Golden Plover association. Ornis Fennica 60, 10–15.

Caldwell, G.S., 1981. Attraction to tropical mixed-species heron flocks: proximate mechanism and consequences. Behavioral Ecology and Sociobiology 8, 99–103.

Caldwell, G.S., 1986. Predation as a selective force on foraging herons: effects of plumage color and flocking. Auk 103, 494–505.

Callaway, R.M., 1995. Positive interactions among plants. Botanical Review 61, 306–349.

Camacho-Cervantes, M., Ojanguren, A.F., Deacon, A.E., Ramnarine, I.W., Maguran, A.E., 2013. Association tendency and preference for heterospecifics in an invasive species. Behaviour 151, 769–780.

Camphuysen, C.J., Webb, A., 1999. Multi-species feeding associations in North Sea seabirds: jointly exploiting a patchy environment. Ardea 87, 177–198.

Campobello, D., Sarà, M., Hare, J.F., 2011. Under my wing: lesser kestrels and jackdaws derive reciprocal benefits in mixed-species colonies. Behavioral Ecology 23, 425–433.

Canales-Delgadillo, J.C., Scott-Morales, L.M., Correa, M.C., Moreno, M.P., 2008. Observations on flocking behavior of Worthen's sparrow (*Spizella wortheni*) and occurrence in mixed-species flocks. Wilson Journal of Ornithology 120, 569–574.

Carbone, C., DuToit, J.T., Gordon, I.J., 1997. Feeding success in African wild dogs: does kleptoparasitism by spotted hyenas influence hunting group size? Journal of Animal Ecology 66, 318–326.

Caro, T.M., 2005. Antipredator Defenses in Birds and Mammals. University of Chicago Press, Chicago.

Caro, T.M., Girling, S., 2010. Conservation by Proxy: Indicator, Umbrella, Keystone, Flagship, and Other Surrogate Species. Island Press, Washington, DC.

Cary, M., 1901. Birds of the Black Hills. Auk 18, 231–238.

Cestari, C., 2009. Heterospecific sociality of birds on beaches from southeastern Brazil. Zoologia 26, 594–600.

Chapman, C.A., Chapman, L.J., 2000. Interdemic variation in mixed-species association patterns: common diurnal primates of Kibale National Park, Uganda. Behavioral Ecology and Sociobiology 47, 129–139.

Chaves-Campos, J., 2003. Localization of army-ant swarms by ant-following birds on the Caribbean slope of Costa Rica: following the vocalization of ant birds to find the swarms. Ornithologia Neotropical 14, 289–294.

Chen, C.C., Hsieh, F., 2002. Composition and foraging behaviour of mixed-species flocks led by the Grey-cheeked Fulvetta in Fushan Experimental Forest, Taiwan. Ibis 144, 317–330.

Chilton, G., Sealy, S.G., 1987. Species roles in mixed-species feeding flocks of seabirds. Journal of Field Ornithology 58, 456–465.

Christman, G.M., 1957. Some interspecific relations in the feeding of estuarine birds. Condor 59, 343.

Chu, M., 2001. Heterospecific responses to scream calls and vocal mimicry by phainopeplas (*Phainopepla nitens*) in distress. Behaviour 138, 775–787.

Clark, K.L., Robertson, R.J., 1979. Spatial and temporal multi-species nesting aggregations in birds as anti-parasite and anti-predator defenses. Behavioral Ecology and Sociobiology 5, 359–371.

Clergeau, P., 1990. Mixed flocks feeding with starlings: an experimental field study in western Europe. Bird Behavior 8, 95–100.

Clua, É., Grosvalet, F., 2001. Mixed-species feeding aggregation of dolphins, large tunas and seabirds in the Azores. Aquatic Living Resources 14, 11–18.

Cockburn, A., 1998. Evolution of helping behavior in cooperatively breeding birds. Annual Review of Ecology and Systematics 29, 141–177.

Cody, M.L., 1971. Finch flocks in the Mohave Desert. Theoretical Population Biology 2, 142–158.

Colorado, G.J., 2013. Why animals come together, with the special case of mixed-species bird flocks. Revista EIA Escuela de Ingeniería de Antioquia 10, 49–66.

Colorado, G.J., Rodewald, A.D., 2015. Response of mixed-species flocks to habitat alteration and deforestation in the Andes. Biological Conservation 188, 72–81.

Conner, R.N., Prather, I.D., Adkisson, C.S., 1975. Common raven and starling reliance on sentinel common crow. Condor 77, 517.

Contreras, T.A., Sieving, K.E., 2011. Leadership of winter mixed-species flocks by tufted titmice (*Baeolophus bicolor*): are titmice passive nuclear species? International Journal of Zoology 2011, 670548.

Coolen, I., van Bergen, Y., Day, R.L., Laland, K.N., 2003. Species difference in adaptive use of public information in sticklebacks. Proceedings of the Royal Society of London B: Biological Sciences 270, 2413–2419.

Cordeiro, N.J., Borghesio, L., Joho, M.P., Monoski, T.J., Mkongewa, V.J., Dampf, C.J., 2015. Forest fragmentation in an African biodiversity hotspot impacts mixed-species bird flocks. Biological Conservation 188, 61–71.

Cords, M., 1990. Mixed-species association of East African guenons: general patterns or specific examples? American Journal of Primatology 21, 101–114.

Cords, M., 2000. Mixed-species association and group movement. In: Boinski, S., Gerber, P.A. (Eds.), On the Move: How and Why Animals Travel in Groups. University of Chicago Press, Chicago, pp. 73–99.

Cords, M., Würsig, B., 2014. A mix of species: associations of heterospecifics among primates and dolphins. In: Yamagiwa, J., Karczmarski, L. (Eds.), Primates and Cetaceans: Field Research and Conservation of Complex Mammalian Societies. Springer Japan, Tokyo.

Corlett, R.T., 2015. The Anthropocene concept in ecology and conservation. Trends in Ecology and Evolution 30, 36–41.

Côté, I.M., 2000. Evolution and ecology of cleaning symbioses in the sea. Oceanography and Marine Biology 38, 311–355.

Côté, I.M., Cheney, K.L., 2005. Animal mimicry: choosing when to be a cleaner-fish mimic. Nature 433, 211–212.

Coulson, G., 1999. Monospecific and heterospecific grouping and feeding behavior in grey kangaroos and red-necked wallabies. Journal of Mammalogy 80, 270–282.

Courser, W.D., Dinsmore, J.J., 1975. Foraging associates of White Ibis. Auk 92, 599–601.

Croft, D.P., James, R., Krause, J., 2008. Exploring Animal Social Networks. Princeton University Press, Princeton.

Croxall, J.P., 1976. The composition and behaviour of some mixed-species bird flocks in Sarawak. Ibis 118, 333–346.

Curio, E., 1978. The adaptive significance of avian mobbing: I. Teleonomic hypotheses and predictions. Zeitschrift für Tierpsychologie 48, 175–183.

Cuthill, I.C., 2006. Color perception. In: Hill, G.E., McGraw, K.J. (Eds.), Bird Coloration. Harvard University Press, Cambridge. MA, pp. 3–40.

Dafni, J., Diamant, A., 1984. School-oriented mimicry: a new type of mimicry in fish. Marine Ecology Progress Series 20, 45–50.

Dall, S.R.X., Giraldeau, L.-A., Olsson, O., McNamara, J.M., Stephens, D.W., 2005. Information and its use by animals in evolutionary ecology. Trends in Ecology and Evolution 20, 187–193.

Danchin, E., Giraldeau, L.-A., Valone, T.J., Wagner, R.H., 2004. Public information: from nosy neighbors to cultural evolution. Science 305, 487–491.

Dargent, F., Torres-Dowdall, J., Scott, M.E., Ramnarine, I., Fussmann, G.F., 2013. Can mixed-species groups reduce individual parasite load? A field test with two closely related Poeciliid fishes (*Poecilia reticulata* and *Poecilia picta*). PLOS ONE 8, e56789.

Darrah, A.J., Smith, K.G., 2013. Comparison of foraging behaviors and movement patterns of the wedge-billed woodpecker (*Glyphorynchus spirurus*) traveling alone and in mixed-species flocks in Amazonian Ecuador. Auk 130, 629–636.

Davis, D.E., 1946. A seasonal analysis of mixed flocks of birds in Brazil. Ecology 27, 168–181.

Dawkins, R., 1982. The Extended Phenotype: The Gene as the Unit of Selection. Freeman, Oxford.

Dawson, E.H., Chittka, L., 2012. Conspecific and heterospecific information use in bumblebees. PLOS ONE 7, e31444.

De Wert, L., Mahon, K., Ruxton, G.D., 2012. Protection by association: evidence for aposematic commensalism. Biological Journal of the Linnean Society 106, 81–89.

Dean, S., 1990. Composition and seasonality of mixed-species flocks of insectivorous birds. Notornis 37, 27–36.

Delamain, J., 1933. Why Birds Sing. Coward-McCann, New York.

Della-Flora, F., Melo, G.L., Sponchiado, J., Cáceres, N.C., 2013. Association of the southern Amazon red squirrel *Urosciurus spadiceus* Olfers, 1818 with mixed-species bird flocks. Mammalia 77, 113–117.

Develey, P.F., Peres, C.A., 2000. Resource seasonality and the structure of mixed species bird flocks in a coastal Atlantic forest of southeastern Brazil. Journal of Tropical Ecology 16, 33–53.

Develey, P.F., Stouffer, P.C., 2001. Effects of roads on movements by understorey birds in mixed-species flocks in central Amazonian Brazil. Conservation Biology 15, 1416–1422.

Diamond, J., 1987. Flocks of brown and black New Guinean birds: a bicolored mixed-species foraging association. Emu 87, 201–211.

Diamond, J.M., 1981. Mixed-species foraging groups. Nature 292, 408–409.

Diamond, J.M., 1982. Mimicry of friarbirds by orioles. Auk 99, 187–196.

Dinsmore, J.J., 1973. Foraging success of Cattle Egrets, *Bubulcus ibis*. American Midland Naturalist 89, 242–246.

Dolby, A.S., Grubb, T.C., 1999. Functional roles in mixed-species foraging flocks: a field manipulation. Auk 116, 557–559.

Dolby, A.S., Grubb Jr., T.C., 1998. Benefits to satellite members in mixed-species foraging groups: an experimental analysis. Animal Behaviour 56, 501–509.

Dolby, A.S., Grubb Jr., T.C., 2000. Social context affects risk taking by a satellite species in a mixed-species foraging group. Behavioral Ecology 11, 110–114.

Dominey, W.J., 1983. Mobbing in colonially nesting fishes, especially the bluegill, *Lepomis macrochirus*. Copeia 1983, 1086–1088.

Duffy, D.C., 1983. The foraging ecology of Peruvian seabirds. Auk 100, 800–810.

Duffy, D.C., 1989. Seabird foraging aggregations: a comparison of two southern upwellings. Colonial Waterbirds 12, 164–175.

Dunn, R.R., Harris, N.C., Colwell, R.K., Koh, L.P., Sodhi, N.S., 2009. The sixth mass coextinction: are most endangered species parasites and mutualists? Proceedings of the Royal Society of London B: Biological Sciences 276, 3037–3045.

Eaton, R.L., 1969. Cooperative hunting by cheetahs and jackals and a theory of domestication of the dog. Mammalia 33, 87–92.

Eaton, S.W., 1953. Wood-warblers wintering in Cuba. Wilson Bulletin 65, 169–174.

Eckardt, W., Zuberbühler, K., 2004. Cooperation and competition in two forest monkeys. Behavioral Ecology 15, 400–411.

Eckburg, P.B., Bik, E.M., Bernstein, C.N., Purdom, E., Dethlefsen, L., Sargent, M., Gill, S.R., Nelson, K.E., Relman, D.A., 2005. Diversity of the human intestinal microbial flora. Science 308, 1635–1638.

Eguchi, K., Yamagishi, S., Randrianasolo, V., 1993. The composition and foraging behaviour of mixed-species flocks of forest-living birds in Madagascar. Ibis 135, 91–96.

Ehrlich, P.R., Ehrlich, A.H., 1973. Coevolution: heterospecific shoaling in Caribbean reef fishes. American Naturalist 107, 157–160.

Eiserer, L.A., 1984. Communal roosting in birds. Bird Behavior 5, 61–80.

Elgar, M.A., 1994. Experimental evidence of a mutualistic association between two web-building spiders. Journal of Animal Ecology 63, 880–886.

Elliott, D.G., 1913. A Review of the Primates. American Museum of Natural History, New York.

Emlen, S.T., Ambrose, H.W., 1970. Feeding interactions of Snowy Egrets and Red-breasted Mergansers. Auk 87, 164–165.

Erisman, B., Heyman, W., Kobara, S., Ezer, T., Pittman, S., Aburto-Oropeza, O., Nemeth, R.S., 2015. Fish spawning aggregations: where well-placed management actions can yield big benefits for fisheries and conservation. Fish and Fisheries 2015, 12132.

Erwin, R.M., 1983. Feeding habitats of nesting wading birds: spatial use and social influences. Auk 100, 960–970.

Ewert, D.N., Askins, R.A., 1991. Flocking behavior of migratory warblers in winter in the Virgin Islands. Condor 93, 864–868.

Farine, D.R., Aplin, L.M., Sheldon, B.C., Hoppitt, W., 2015a. Interspecific social networks promote information transmission in wild songbirds. Proceedings of the Royal Society of London B: Biological Sciences 282, 20142804.

Farine, D.R., Downing, C.P., Downing, P.A., 2014. Mixed-species associations can arise without heterospecific attraction. Behavioral Ecology 25, 574–581.

Farine, D.R., Garroway, C.J., Sheldon, B.C., 2012. Social network analysis of mixed-species flocks: Exploring the structure and evolution of interspecific social behaviour. Animal Behaviour 84, 1271–1277.

Farine, D.R., Milburn, P.J., 2013. Social organisation of thornbill-dominated mixed-species flocks using social network analysis. Behavioral Ecology and Sociobiology 67, 321–330.

Farine, D.R., Montiglio, P.O., Spiegel, O., 2015b. From individuals to groups and back: the evolutionary implications of group phenotypic composition. Trends in Ecology and Evolution 30, 609–621.

Farley, E.A., Sieving, K.E., Contreras, T.A., 2008. Characterizing complex mixed-species bird flocks using an objective method for determining species' participation. Journal of Ornithology 149, 451–468.

Fautin, D.G., 1991. The anemonefish symbiosis: what is known and what is not. Symbiosis 10, 23–46.

Feldman, T.S., Morris, W.F., Wilson, W.G., 2004. When can two plant species facilitate each other's pollination? Oikos 105, 197–207.

Fenwick, G.D., 1978. Plankton swarms and their predators at the Snares Islands. New Zealand Journal of Marine and Freshwater Research 12, 223–224.

Fernández-Juricic, E., 2000. Forest fragmentation affects winter flock formation of an insectivorous guild. Ardea 88, 235–241.

Fernández-Juricic, E., 2002. Nested patterns of species distribution and winter flock occurrence of insectivorous birds in a fragmented landscape. Ecoscience 9, 450–458.

Fernandez, E.V., Li, Z., Zheng, W., Ding, Y., Sun, D., Che, Y., 2014. Intraspecific host selection of Père David's deer by cattle egrets in Dafeng, China. Behavioural Processes 105, 36–39.

Ferrari, M.C.O., Chivers, D.P., 2008. Cultural learning of predator recognition in mixed-species assemblages of frogs: the effect of tutor-to-observer ratio. Animal Behaviour 75, 1921–1925.

Ferrari, M.C.O., Wisenden, B.D., Chivers, D.P., 2010. Chemical ecology of predator-prey interactions in aquatic ecosystems: a review and prospectus. Canadian Journal of Zoology 88, 698–724.

FitzGibbon, C.D., 1990. Mixed-species grouping in Thomson's and Grant's gazelles: the antipredator benefits. Animal Behaviour 39, 1116–1126.

FitzGibbon, C.D., 1994. The costs and benefits of predator inspection behaviour in Thomson's gazelles. Behavioral Ecology and Sociobiology 34, 139–148.

Fleury, M.C., Gautier-Hion, A., 1997. Better to live with allogenerics than to live alone? The case of single male *Cercopithecus pogonias* in troops of *Colobus satanas*. International Journal of Primatology 18, 967–974.

Flower, T., 2011. Fork-tailed drongos use deceptive mimicked alarm calls to steal food. Proceedings of the Royal Society of London B: Biological Sciences 278, 1548–1555.

Flower, T.P., Child, M.F., Ridley, A.R., 2013. The ecological economics of kleptoparasitism: payoffs from self-foraging versus kleptoparasitism. Journal of Animal Ecology 82, 245–255.

Flower, T.P., Gribble, M., Ridley, A.R., 2014. Deception by flexible alarm mimicry in an African bird. Science 344, 513–516.

Forsman, J.T., Martin, T.E., 2009. Habitat selection for parasite-free space by hosts of parasitic cowbirds. Oikos 118, 464–470.

Frantzis, A., Herzing, D.L., 2002. Mixed-species associations of striped dolphins (*Stenella coeruleoalba*), short-beaked common dolphins (*Delphinus delphis*), and Risso's dolphins (*Grampus griseus*) in the Gulf of Corinth (Greece, Mediterranean Sea). Aquatic Mammals 28, 188–197.

Freed, B.Z., 2007. Polyspecific associations of Crowned Lemurs and Sanford's Lemurs in Madagascar. In: Gould, L., Sauther, M.L. (Eds.), Lemurs. Springer, New York, pp. 111–132.

French, A.R., Smith, T.B., 2005. Importance of body size in determining dominance hierarchies among diverse tropical frugivores. Biotropica 37, 96–101.

Fuong, H., Keeley, K.N., Bulut, Y., Blumstein, D.T., 2014. Heterospecific alarm call eavesdropping in nonvocal, white-bellied copper-striped skinks, *Emoia cyanura*. Animal Behaviour 95, 129–135.

Galef, B.G., Giraldeau, L.A., 2001. Social influences on foraging in vertebrates: causal mechanisms and adaptive functions. Animal Behaviour 61, 3–15.

Gannon, G.R., 1934. Associations of small insectivorous birds. Emu 34, 122–129.

Garber, P.A., 1988. Diet, foraging patterns, and resource defense in a mixed species troop of *Saguinus mystax* and *Saguinus fuscicollis* in Amazonian Peru. Behaviour 105, 18–34.

Gartlan, J.S., Struhsaker, T.T., 1972. Polyspecific associations and niche separation of rain-forest anthropoids in Cameroon, West Africa. Journal of Zoology 168, 221–265.

Gautier-Hion, A., Quris, R., Gautier, J.-P., 1983. Monospecific vs polyspecific life: a comparative study of foraging and antipredatory tactics in a community of *Cercopithecus* monkeys. Behavioral Ecology and Sociobiology 12, 325–335.

Gautier-Hion, A., Tutin, C.E.G., 1988. Simultaneous attack by adult males of a polyspecific troop of monkeys against a crowned hawk eagle. Folia Primatologica 51, 149–151.

Gavrilov, V.V., 2015. Advantages and disdvantages of interspecies associations of migratory northern sandpipers (Charadrii, Aves). Biology Bulletin Reviews 5, 63–70.

Gibson, L., Lee, T.M., Koh, L.P., Brook, B.W., Gardner, T.A., Barlow, J., Peres, C.A., Bradshaw, C.J.A., Laurance, W.F., Lovejoy, T.E., Sodhi, N.S., 2011. Primary forests are irreplaceable for sustaining tropical biodiversity. Nature 478, 378–383.

Gibson, R.M., Aspbury, A.S., McDaniel, L.L., 2002. Active formation of mixed–species grouse leks: a role for predation in lek evolution? Proceedings of the Royal Society of London B: Biological Sciences 269, 2503–2507.

Gillies, C.S., St Clair, C.C., 2008. Riparian corridors enhance movement of a forest specialist bird in fragmented tropical forest. Proceedings of the National Academy of Sciences of the United States of America 105, 19774–19779.

Giraldeau, L.-A., Caraco, T., 2000. Social Foraging Theory. Princeton University Press, Princeton.

Glos, J., Dausmann, K.H., Linsenmair, E.K., 2007a. Mixed-species social aggregations in Madagascan tadpoles — determinants and species composition. Journal of Natural History 41, 1965–1977.

Glos, J., Erdmann, G., Dausmann, K.H., Linsenmair, K.E., 2007b. A comparative study of predator-induced social aggregation of tadpoles in two anuran species from western Madagascar. The Herpetological Journal 17, 261–268.

Godin, J.-G., Davis, J.and S.A., 1995. Who dares, benefits: predator approach behaviour in the guppy (*Poecilia reticulata*) deters predator pursuit. Proceedings of the Royal Society of London B: Biological Sciences 259, 193–200.

Goldsworthy, S.D., Boness, D.J., Fleischer, R.C., 1999. Mate choice among sympatric fur seals: female preference for conphenotypic males. Behavioral Ecology and Sociobiology 45, 253–267.

Gómez, J.P., Bravo, G.A., Brumfield, R.T., Tello, J.G., Cadena, C.D., 2010. A phylogenetic approach to disentangling the role of competition and habitat filtering in community assembly of neotropical forest birds. Journal of Animal Ecology 79, 1181–1192.

Gonzalez, A.D., Matta, N.E., Ellis, V.A., Miller, E.T., Ricklefs, R.E., Gutierrez, H.R., 2014. Mixed species flock, nest height, and elevation partially explain avian haemoparasite prevalence in Colombia. PLOS ONE 9, e100695.

Goodale, E., Beauchamp, G., 2010. The relationship between leadership and gregariousness in mixed-species bird flocks. Journal of Avian Biology 41, 99–103.

Goodale, E., Beauchamp, G., Magrath, R.D., Nieh, J.C., Ruxton, G.D., 2010. Interspecific information transfer influences animal community structure. Trends in Ecology and Evolution 25, 354–361.

Goodale, E., Ding, P., Liu, X., Martínez, A., Walters, M., Robinson, S.K., 2015. The structure of multi-species flocks and their role in the organization of forest bird communities, with special reference to China. Avian Research 6, 14.

Goodale, E., Goodale, U., Mana, R., 2012. The role of toxic pitohuis in mixed-species flocks of lowland forest in Papua New Guinea. Emu 112, 9–16.

Goodale, E., Kotagama, S.W., 2005a. Alarm calling in Sri Lankan mixed-species bird flocks. Auk 122, 108–120.

Goodale, E., Kotagama, S.W., 2005b. Testing the roles of species in mixed-species bird flocks of a Sri Lankan rainforest. Journal of Tropical Ecology 21, 669–676.

Goodale, E., Kotagama, S.W., 2006. Vocal mimicry by a passerine bird attracts other species involved in mixed-species flocks. Animal Behaviour 72, 471–477.

Goodale, E., Kotagama, S.W., 2008. Response to conspecific and heterospecific alarm calls in mixed-species bird flocks of a Sri Lankan rainforest. Behavioral Ecology 19, 887–894.

Goodale, E., Nieh, J.C., 2012. Public use of olfactory information associated with predation in two species of social bees. Animal Behaviour 84, 919–924.

Goodale, E., Nizam, B.Z., Robin, V.V., Sridhar, H., Trivedi, P., Kotagama, S.W., Padmalal, U.K.G.K., Perera, R., Pramod, P., Vijayan, L., 2009. Regional variation in the composition and structure of mixed-species bird flocks in the Western Ghats and Sri Lanka. Current Science 97, 648–663.

Goodale, E., Ratnayake, C.P., Kotagama, S.W., 2014. Vocal mimicry of alarm-associated sounds by a drongo elicits flee and mobbing responses from other species that participate in mixed-species bird flocks. Ethology 120, 266–274.

Gosling, L.M., 1980. Defense guilds of savannah ungulates as a context for scent communication. Symposium of the Zoological Society of London 45, 195–212.

Goulson, D., Hawson, S.A., Stout, J.C., 1998. Foraging bumblebees avoid flowers already visited by conspecifics or by other bumblebee species. Animal Behaviour 55, 199–206.

Gowans, S., Würsig, B., Karczmarski, L., 2008. The social structure and strategies of delphinids: predictions based on an ecological framework. Advances in Marine Biology 53, 195–294.

Grade, A.M., Sieving, K.E., 2016. When the birds go unheard: highway noise disrupts information transfer between bird species. Biology Letters 12, 20160113.

Gradwohl, J., Greenberg, R., 1980. The formation of antwren flocks on Barro Colorado Island, Panama. Auk 97, 385–395.

Gram, W.K., 1998. Winter participation by Neotropical migrant and resident birds in mixed-species flocks in northeastern Mexico. Condor 100, 44–53.

Graves, G.R., Gotelli, N.J., 1993. Assembly of avian mixed-species flocks in Amazonia. Proceedings of the National Academy of Sciences of the United States of America 90, 1388–1391.

Green, M.C., Leberg, P.L., 2005. Flock formation and the role of plumage colouration in Ardeidae. Canadian Journal of Zoology 83, 683–693.

Greenberg, R., 2000. Birds of many feathers: the formation and structure of mixed-species flocks of forest birds. In: Boinski, S., Garber, P.A. (Eds.), On the Move: How and Why Animals Travel in Groups. University of Chicago Press, Chicago, pp. 521–558.

Greig-Smith, P.W., 1981. The role of alarm responses in the formation of mixed-species flocks of heathland birds. Behavioral Ecology and Sociobiology 8, 7–10.

Griffin, A.S., Savani, R.S., Hausmanis, K., Lefebvre, L., 2005. Mixed-species aggregations in birds: zenaida doves, *Zenaida aurita*, respond to the alarm calls of carib grackles, *Quiscalus lugubris*. Animal Behaviour 70, 507–515.

Groom, M.J., 1992. Sand-colored Nighthawks parasitize the antipredator behavior of three nesting bird species. Ecology 73, 785–793.

Grostal, P., Walter, D.E., 1997. Kleptoparasites or commensals? Effects of *Argyrodes antipodianus* (Araneae: Theridiidae) on *Nephila plumipes* (Araneae: Tetragnathidae). Oecologia 111, 570–574.

Grover, J.J., Olla, B.L., 1983. The role of the Rhinoceros Auklet (*Cerorhinca monocerata*) in mixed-species feeding assemblages of seabirds in the Strait of Juan de Fuca, Washington. Auk 100, 979–982.

Grutter, A.S., 1999. Cleaner fish really do clean. Nature 398, 672–673.

Grutter, A.S., Murphy, J.M., Choat, J.H., 2003. Cleaner fish drives local fish diversity on coral reefs. Current Biology 13, 64–67.

Gygax, L., 2002. Evolution of group size in the dolphins and porpoises: interspecific consistency of intraspecific patterns. Behavioral Ecology 13, 583–590.

Gyimesi, A., van Lith, B., Nolet, B.A., 2012. Commensal foraging with Bewick's Swans *Cygnus bewickii* doubles instantaneous intake rate of Common Pochards *Aythya ferina*. Ardea 100, 55–62.

Haddad, N.M., Brudvig, L.A., Clobert, J., Davies, K.F., Gonzalez, A., Holt, R.D., Lovejoy, T.E., Sexton, J.O., Austin, M.P., Collins, C.D., Cook, W.M., Damschen, E.I., Ewers, R.M., Foster, B.L., Jenkins, C.N., King, A.J., Laurance, W.F., Levey, D.J., Margules, C.R., Melbourne, B.A., Nicholls, A.O., Orrock, J.L., Song, D.-X., Townshend, J.R., 2015. Habitat fragmentation and its lasting impact on Earth's ecosystems. Science Advances 1, e1500052.

Haemig, P.D., 2001. Symbiotic nesting of birds with formidable animals: a review with applications to biodiversity conservation. Biodiversity and Conservation 10, 527–540.

Hamilton, W.D., 1971. Geometry for the selfish herd. Journal of Theoretical Biology 31, 295–311.

Haney, J.C., Fristrup, K.M., Lee, D.S., 1992. Geometry of visual recruitment by seabirds to ephemeral foraging flocks. Ornis Scandinavica 23, 49–62.

Harrison, N.M., Whitehouse, M.J., 2011. Mixed-species flocks: an example of niche construction? Animal Behaviour 81, 675–682.

Harrison, N.M., Whitehouse, M.J., Heinemann, D., Prince, P.A., Hunt, G.L., Veit, R.R., 1991. Observations of multispecies seabird flocks around South Georgia. Auk 108, 801–810.

Hay, M.E., 1986. Associational plant defenses and the maintenance of species diversity: turning competitors into accomplices. American Naturalist 128, 617–641.

Heinrich, B., Vogt, F.D., 1980. Aggregation and foraging behavior of whirligig beetles (Gyrinidae). Behavioral Ecology and Sociobiology 7, 179–186.

Hénaut, Y., Machkour-M'Rabet, S., 2010. Interspecific aggregation around the web of the orb spider *Nephila clavipes*: consequences for the web architecture of *Leucauge venusta*. Ethology Ecology and Evolution 22, 203–209.

Herremans, M., 1990. Can night migrants use interspecific song recognition to assess habitat? Gerfaut 80, 141–148.

Herzing, D.L., Elliser, C.R., 2013. Directionality of sexual activities during mixed-species encounters between Atlantic spotted dolphin (*Stenella frontalis*) and bottlenose dolphins (*Tursops truncatus*). International Journal of Comparative Psychology 26, 124–134.

Herzing, D.L., Johnson, C.M., 1997. Interspecific interactions between Atlantic spotted dolphins (*Stenella frontalis*) and bottlenose dolphins (*Tursiops truncatus*) in the Bahamas, 1985-1995. Aquatic Mammals 23, 85–99.

Herzing, D.L., Moewe, K., Brunnick, B.J., 2003. Interspecies interactions between Atlantic spotted dolphins, *Stenella frontalis*, and bottlenose dolphins, *Tursiops truncatus*, on Great Bahama Bank, Bahamas. Aquatic Mammals 29, 335–341.

Hetrick, S.A., Sieving, K.E., 2011. Antipredator calls of tufted titmice and interspecific transfer of encoded threat information. Behavioral Ecology 23 (1), 83–92.

Heymann, E.W., 1997. The relationship between body size and mixed-species troops of tamarins (*Saguinus* spp.). Folia Primatologica 68, 287–295.

Heymann, E.W., Buchanan-Smith, H.M., 2000. The behavioural ecology of mixed-species troops of callitricine primates. Biological Reviews 75, 169–190.

Heymann, E.W., Hsia, S.S., 2015. Unlike fellows – a review of primate–non-primate associations. Biological Reviews 90, 142–156.

Hindwood, K.A., 1937. The flocking of birds with particular reference to the association of small insectivorous birds. Emu 36, 254–261.

Hino, T., 1998. Mutualistic and commensal organization of avian mixed-species foraging flocks in a forest of western Madagascar. Journal of Avian Biology 29, 17–24.

Hino, T., 2000. Intraspecific differences in benefits from feeding in mixed-species flocks. Journal of Avian Biology 31, 441–446.

Hodge, M.A., Uetz, G.W., 1996. Foraging advantages of mixed-species association between solitary and colonial orb-weaving spiders. Oecologia 107, 578–587.

Hodgins, N.K., Dolman, S.J., Weir, C.R., 2014. Potential hybridism between free-ranging Risso's dolphins (*Grampus griseus*) and bottlenose dolphins (*Tursiops truncatus*) off north-east Lewis (Hebrides, UK). Marine Biodiversity Records 7, e97.

Hoffman, W., Heinemann, D., Wiens, J.A., 1981. The ecology of seabird feeding flocks in Alaska. Auk 98, 437–456.

Höglund, J., Alatalo, R.V., 1995. Leks. Princeton University Press, Princeton.

Holenweg, A.-K., Noë, R., Schabel, M., 1996. Waser's gas model applied to associations between red colobus and Diana monkeys in the Tai National Park, Ivory Coast. Folia Primatologica 67, 125–136.

Hudson, A.V., Furness, R.W., 1988. Utilization of discarded fish by scavenging seabirds behind whitefish trawlers in Shetland. Journal of Zoology 215, 151–166.

Hughes, D.P., Pierce, N.E., Boomsma, J.J., 2008. Social insect symbionts: evolution in homeostatic fortresses. Trends in Ecology and Evolution 23, 672–677.

Hunt, G.L., Harrison, N.M., Hamner, W.M., Obst, B.S., 1988. Observations of a mixed-species flock of birds foraging on euphausiids near St. Matthew Island, Bering Sea. Auk 105, 345–349.

Hurd, C.R., 1996. Interspecific attraction to the mobbing calls of black-capped chickadees (*Parus atricapillus*). Behavioral Ecology and Sociobiology 38, 287–292.

Hurlbert, S.H., Lopez, M., Keith, J.O., 1984. Wilson's phalarope in the Central Andes and its interaction with the Chilean flamingo. Revista Chilena de Historia Natural 57, 47–57.

Hutto, R.L., 1987. A description of mixed-species insectivorous bird flocks in western Mexico. Condor 89, 282–292.

Hutto, R.L., 1988. Foraging behaviour patterns suggest a possible cost associated with participation in mixed-species bird flocks. Oikos 51, 79–83.

Hutto, R.L., 1994. The composition and social organization of mixed-species flocks in a tropical deciduous forest in western Mexico. Condor 96, 105–118.

Ioannou, C.C., Krause, J., 2008. Searching for prey: the effects of group size and number. Animal Behaviour 75, 1383–1388.

Isack, H.A., Reyer, H.-U., 1989. Honeyguides and honey gatherers: interspecific communication in a symbiotic relationship. Science 243, 1343–1346.

Jackson, J.B.C., Kirby, M.X., Berger, W.H., Bjorndal, K.A., Botsford, L.W., Bourque, B.J., Bradbury, R.H., Cooke, R., Erlandson, J., Estes, J.A., Hughes, T.P., Kidwell, S., Lange, C.B., Lenihan, H.S., Pandolfi, J.M., Peterson, C.H., Steneck, R.S., Tegner, M.J., Warner, R.R., 2001. Historical overfishing and the recent collapse of coastal ecosystems. Science 293, 629–638.

Jacobsen, O.W., Ugelvik, M., 1994. Effects of waders on grazing and vigilance behaviour in breeding wigeon, *Anas penelope*. Animal Behaviour 47, 488–490.

Janzen, D.H., 1985. The natural history of mutualisms. In: Boucher, D.H. (Ed.), The Biology of Mutualism: Ecology and Evolution. Croom Helm, London, pp. 40–99.

Jayarathna, A., Kotagama, S.W., Goodale, E., 2013. The seasonality of mixed-species bird flocks in a Sri Lankan rainforest in relation to the breeding of the nuclear species, Orange-billed Babbler *Turdoides rufescens*. Forktail 29, 138–139.

Johnson, F.R., McNaughton, E.J., Shelley, C.D., Blumstein, D.T., 2003. Mechanisms of heterospecific recognition in avian mobbing calls. Australian Journal of Zoology 51, 577–585.

Jolles, J.W., King, A.J., Manica, A., Thornton, A., 2013. Heterogeneous structure in mixed-species corvid flocks in flight. Animal Behaviour 85, 743–750.

Jones, S.E., 1977. Coexistence in mixed species antwren flocks. Oikos 28, 366–375.

Jonsson, J.E., Afton, A.D., 2009. Time budgets of Snow Geese *Chen caerulescens* and Ross's Geese *Chen rossii* in mixed flocks: implications of body size, ambient temperature and family associations. Ibis 151, 134–144.

Joseph, L.N., Maloney, R.F., Possingham, H.P., 2009. Optimal allocation of resources among threatened species: a project prioritization protocol. Conservation Biology 23, 328–338.

Jullien, M., Clobert, J., 2000. The survival value of flocking in neotropical birds: reality or fiction? Ecology 81, 3416–3430.

Jullien, M., Thiollay, J.-M., 1998. Multi-species territoriality and dynamic of neotropical forest understorey bird flocks. Journal of Animal Ecology 67, 227–252.

Källander, H., 1993. Commensal feeding associations between Yellow Wagtails *Motacilla flava* and cattle. Ibis 135, 97–100.

Källander, H., 2005. Commensal association of waterfowl with feeding swans. Waterbirds 28, 326–330.

Kane, A., Jackson, A.L., Ogada, D.L., Monadjem, A., McNally, L., 2014. Vultures acquire information on carcass location from scavenging eagles. Proceedings of the Royal Society of London B: Biological Sciences 281.

Karban, R., 2008. Plant behaviour and communication. Ecology Letters 11, 727–739.

Karplus, I., Thompson, A.R., 2011. The partnership between gobiid fishes and burrowing alpheid shrimps. In: Patzner, R., Van Tassell, J.L., Kovacic, M., Kapoor, B.G. (Eds.), Biology of Gobies. CRC Press, New Hampshire, pp. 559–608.

Kelley, L.A., Healy, S.D., 2012. Vocal mimicry in spotted bowerbirds is associated with an alarming context. Journal of Avian Biology 43, 525–530.

Kent, D.I., Fisher, J.D., Marliave, J.B., 2011. Interspecific nesting in marine fishes: spawning of the spinynose sculpin, *Asemichthys taylori*, on the eggs of the buffalo sculpin, *Enophrys bison*. Ichthyological Research 58, 355–359.

Kern, J.M., Radford, A.N., 2013. Call of duty? Variation in use of the watchman's song by sentinel dwarf mongooses, *Helogale parvula*. Animal Behaviour 85, 967–975.

Kiffner, C., Kioko, J., Leweri, C., Krause, S., 2014. Seasonal patterns of mixed species groups in large East African mammals. PLOS ONE 9, e113446.

King, D.I., Rappole, J.H., 2001. Mixed species bird flocks in dipterocarp forest of north-central Burma (Mynamar). Ibis 143, 380–390.

Kiszka, J., 2007. Atypical associations between dugongs (*Dugong dugon*) and dolphins in a tropical lagoon. Journal of the Marine Biological Association of the United Kingdom 87, 101–104.

Kiszka, J., Perrin, W.F., Pusineri, C., Ridoux, V., 2011. What drives island-associated tropical dolphins to form mixed-species associations in the southwest Indian Ocean? Journal of Mammalogy 92, 1105–1111.

Klaus, G., Schmidg, B., 1998. Geophagy at natural licks and mammal ecology: a review. Mammalia 62, 482–498.

Kleinhappel, T.K., Burman, O.H.P., John, E.A., Wilkinson, A., Pike, T.W., 2016. A mechanism mediating inter-individual associations in mixed-species groups. Behavioral Ecology and Sociobiology 70, 755–760.

Kluever, B.M., Howery, L.D., Breck, S.W., Bergman, D.L., 2009. Predator and heterospecific stimuli alter behaviour in cattle. Behavioural Processes 81, 85–91.

Knight, A.T., Cowling, R.M., Rouget, M., Balmford, A., Lombard, A.T., Campbell, B.M., 2008. Knowing but not doing: selecting priority conservation areas and the research–implementation gap. Conservation Biology 22, 610–617.

Knowlton, J.L., Graham, C.H., 2011. Species interactions are disrupted by habitat degradation in the highly threatened Tumbesian region of Ecuador. Ecological Applications 21, 2974–2986.

Kobara, S., Heyman, W.D., Pittman, S.J., Nemeth, R.S., 2013. Biogeography of transient reef-fish spawning aggregations in the Caribbean: a synthesis for future research and management. Oceanography and Marine Biology: An Annual Review 51, 281–325.

Koda, H., 2012. Possible use of heterospecific food-associated calls of macaques by sika deer for foraging efficiency. Behavioural Processes 91, 30–34.

Kotagama, S.W., Goodale, E., 2004. The composition and spatial organization of mixed-species flocks in a Sri Lankan rainforest. Forktail 20, 63–70.

Krajewski, J.P., Bonaldo, R.M., Sazima, C., Sazima, I., 2004. The association of the goatfish *Mulloidichthys martinicus* with the grunt *Haemulon chrysargyreum*: an example of protective mimicry. Biota Neotropica 4, 1–4.

Krajewski, J.P., Bonaldo, R.M., Sazima, C., Sazima, I., 2009. Octopus mimicking its follower reef fish. Journal of Natural History 43, 185–190.

Kramer, D.L., Graham, J.B., 1976. Synchronous air breathing, a social component of respiration in fishes. Copeia 1976, 689–697.

Krams, I., Krama, T., 2002. Interspecific reciprocity explains mobbing behaviour of the breeding chaffinches, *Fringilla coelebs*. Proceedings of the Royal Society of London B: Biological Sciences 269, 2345–2350.

Krause, J., 1993a. The relationship between foraging and shoal position in a mixed shoal of roach (*Rutilus rutilus*) and chub (*Leuciscus cephalus*): a field study. Oecologia 93, 356–359.

Krause, J., 1993b. Transmission of fright reaction between different species of fish. Behaviour 127, 37–48.

Krause, J., Godin, J.-G.J., 1994. Shoal choice in banded killifish: the effects of predation risk, fish size, species composition and size of shoals. Ethology 98, 128–136.

Krause, J., Godin, J.-G.J., Brown, D., 1996. Size-assortiveness in multi-species fish shoals. Journal of Fish Biology 49, 221–225.

Krause, J., Ruxton, G.D., 2002. Living in Groups. Oxford University Press, Oxford.

Krause, J., Ward, A.J.W., Jackson, A.L., Ruxton, G.D., James, R., Currie, S., 2005. The influence of differential swimming speeds on composition of multi-species fish shoals. Journal of Fish Biology 67, 866–872.

Krebs, J.R., 1973. Social learning and the adaptive significance of mixed-species flocks of chickadees. Canadian Journal of Zoology 51, 1275–1288.

Kristiansen, J.N., Fox, A.D., Boyd, H., Stroud, D.A., 2000. Greenland White-fronted Geese *Anser albifrons flavivostris* benefit from feeding in mixed-species flocks. Ibis 142, 142–144.

Kushlan, J.A., 1977. The significance of plumage colour in the formation of feeding aggregations of ciconiiforms. Ibis 119, 361–364.

Kyle, R., 2006. Co-operative feeding by Black Egrets, Little Egrets and African Spoonbills in Ndumo Game Reserve, South Africa. Ostrich 76, 91–92.

Landeau, L., Terborgh, J., 1986. Oddity and the confusion effect in predation. Animal Behaviour 34, 1372–1380.

Larsen, J.K., 1996. Wigeon *Anas penelope* offsetting dependence on water by feeding in mixed-species flocks: a natural experiment. Ibis 138, 555–557.

Latta, S.C., Wunderle, J.M., 1996. The composition and foraging ecology of mixed-species flocks in pine forests of Hispaniola. Condor 98, 595–607.

Laurance, W.F., Koster, H., Grooten, M., Anderson, A.B., Zuidema, P.A., Zwick, S., Zagt, R.J., Lynam, A.J., Linkie, M., Anten, N.P.R., 2012. Making conservation research more relevant for conservation practitioners. Biological Conservation 153, 164–168.

Laurance, W.F., Sayer, J., Cassman, K.G., 2014. Agricultural expansion and its impacts on tropical nature. Trends in Ecology and Evolution 29, 107–116.

Le Guen, R., Corbara, B., Rossi, V., Azémar, F., Dejean, A., 2015. Reciprocal protection from natural enemies in an ant-wasp association. Comptes Rendus Biologies 338, 255–259.

Lea, A.J., Barrera, J.P., Tom, L.M., Blumstein, D.T., 2008. Heterospecific eavesdropping in a non-social species. Behavioral Ecology 19, 1041–1046.

Leclaire, S., Faulkner, C.T., 2014. Gastrointestinal parasites in relation to host traits and group factors in wild meerkats *Suricata suricatta*. Parasitology 141, 925–933.

Lee, A.T.K., Kumar, S., Brightsmith, D.J., Marsden, S.J., 2010. Parrot claylick distribution in South America: do patterns of "where" help answer the question "why"? Ecography 33, 503–513.

Lee, T.M., Soh, M.C.K., Sodhi, N., Koh, L.P., Lim, S.L.-H., 2005. Effects of habitat disturbance on mixed species bird flocks in a tropical sub-montane rainforest. Biological Conservation 122, 193–204.

Li, Z., Jiang, Z., Beauchamp, G., 2010. Nonrandom mixing between groups of Przewalski's gazelle and Tibetan gazelle. Journal of Mammalogy 91, 674–680.

Li, Y.Y., Wang, J., Metzner, W., Luo, B., Jiang, T.L., Yang, S.L., Shi, L.M., Huang, X.B., Yue, X.K., Feng, J., 2014. Behavioral responses to echolocation calls from sympatric heterospecific bats: implications for interspecific competition. Behavioral Ecology and Sociobiology 68, 657–667.

Lima, S.L., 1994. Collective detection of predatory attack by birds in the absence of alarm signals. Journal of Avian Biology 25, 319–326.

Lima, S.L., 1995. Collective detection of predatory attack by social foragers: fraught with ambiguity? Animal Behaviour 50, 1097–1108.

Lindeberg, B., 1964. The swarm of males as a unit for taxonomic recognition in the Chironomids (Diptera). Annales Zoologici Fennici 1, 72–76.

Lönnstedt, O.M., Ferrari, M.C.O., Chivers, D.P., 2014. Lionfish predators use flared fin displays to initiate cooperative hunting. Biology Letters 10, 20140281.

Loukola, O.J., Seppänen, J.-T., Krams, I., Torvinen, S.S., Forsman, J.T., 2013. Observed fitness may affect niche overlap in competing species via selective social information use. American Naturalist 182, 474–483.

Lukoschek, V., McCormick, M.I., 2000. A review of multi-species foraging associations in fishes and their ecological significance. In: Proceedings of the 9th International Coral Reef Symposium. Ministry of Environment, Indonesia; Indonesian Institute of Sciences; International Society for Reef Studies, Bali, pp. 467–474.

Lusseau, D., 2007. Evidence for social role in a dolphin social network. Evolutionary Ecology 21, 357–366.

Madden, J.R., Kilner, R.M., Davies, N.B., 2005. Nestling responses to adult food and alarm calls: 1. species-specific responses in two cowbird hosts. Animal Behaviour 70, 619–627.

Magrath, R.D., Pitcher, B.J., Gardner, J.L., 2009. An avian eavesdropping network: alarm signal reliability and heterospecific response. Behavioral Ecology 20, 745–752.

Magrath, R.D., Haff, T.M., Fallow, P.M., Radford, A.N., 2015a. Eavesdropping on heterospecific alarm calls: from mechanisms to consequences. Biological Reviews 90, 560–586.

Magrath, R.D., Haff, T.M., McLachlan, J.R., Igic, B., 2015b. Wild birds learn to eavesdrop on heterospecific alarm calls. Current Biology 25, 2047–2050.

Mahon, T.E., Kaiser, G.W., Burger, A.E., 1992. The role of Marbled Murrelets in mixed-species feeding flocks in British Columbia. Wilson Bulletin 104, 738–743.

Makenbach, S.A., Waterman, J.M., Roth, J.D., 2013. Predator detection and dilution as benefits of association between yellow mongooses and Cape ground squirrels. Behavioral Ecology and Sociobiology 67, 1187–1194.

Maldonado-Coelho, M., Durães, R., 2003. The Black-googled Tanager (*Trichothraupis melanops*): an occasional kleptoparasite in mixed-species bird flocks and ant swarms of southeastern Brazil. Ornitologia Neotropical 14, 397–403.

Maldonado-Coelho, M., Marini, M.A., 2000. Effects of forest fragment size and successional stage on mixed-species bird flocks in southeastern Brazil. Condor 102, 585–594.

Maldonado-Coelho, M., Marini, M.A., 2004. Mixed-species bird flocks from Brazilian Atlantic forest: the effects of forest fragmentation and seasonality on their size, richness and stability. Biological Conservation 116, 19–26.

Mammides, C., Chen, J., Goodale, U.M., Kotagama, S.W., Sidhu, S., Goodale, E., 2015. Does mixed-species flocking influence how birds respond to land-use intensity? Proceedings of the Royal Society of London B: Biological Sciences 282, 20151118.

Mammides, C., Goodale, U.M., Corlett, R.T., Chen, J., Bawa, K.S., Hariya, H., Jarrad, F., Primack, R.B., Ewing, H., Xia, X., Goodale, E., 2016. Increasing geographic diversity in the international conservation literature: a stalled process? Biological Conservation 198, 78–83.

Manassa, R.P., McCormick, M.I., Chivers, D.P., 2013. Socially acquired predator recognition in complex ecosystems. Behavioral Ecology and Sociobiology 67, 1033–1040.

Maniscalco, J.M., Ostrand, W.D., Suryan, R.M., Irons, D.B., 2001. Passive interference competition by Glaucous-winged Gulls on Black-legged Kittiwakes: a cost of feeding in flocks. Condor 103, 616–619.

Marler, P., 1957. Specific distinctiveness in the communication signals of birds. Behaviour 11, 13–39.

Marras, S., Killen, S.S., Lindstrom, J., McKenzie, D.J., Steffensen, J.F., Domenici, P., 2015. Fish swimming in schools save energy regardless of their spatial position. Behavioral Ecology and Sociobiology 69, 219–226.

Martínez, A.E., Gomez, J.P., 2013. Are mixed-species bird flocks stable through two decades? American Naturalist 181, E53–E59.

Martínez, A.E., Zenil, R.T., 2013. Foraging guild influences dependence on heterospecific alarm calls in Amazonian bird flocks. Behavioral Ecology 23, 544–550.

Martínez, A., Gomez, J.P., Ponciano, J.M., Robinson, S.K., 2016. Functional traits, sociality and perceived predation risk in an Amazonian understory bird community. American Naturalist 187.

Marzluff, J.M., Heinrich, B., 1991. Foraging by common ravens in the presence and absence of territory holders: an experimental analysis of social foraging. Animal Behaviour 42, 755–770.

Marzluff, J.M., Heinrich, B., Marzluff, C.S., 1996. Raven roosts are mobile information centres. Animal Behaviour 51, 89–103.

Master, T.L., 1992. Composition, structure and dynamics of mixed-species foraging aggregations in a southern New Jersey salt marsh. Colonial Waterbirds 15, 66–74.

Mathis, A., Chivers, D.P., 2003. Overriding the oddity effect in mixed-species aggregations: group choice by armoured and non-armoured species. Behavioral Ecology 14, 334–339.

Mathis, A., Chivers, D.P., Smith, R.J.F., 1996. Cultural transmission of predator recognition in fishes: intraspecific and interspecific learning. Animal Behaviour 51, 185–201.

Matthysen, E., Collet, F., Cahill, J., 2008. Mixed flock composition and foraging behavior of insectivorous birds in undisturbed and disturbed fragments of high-Andean Polylepis woodland. Ornitologia Neotropical 19, 403–416.

May-Collado, L.J., 2010. Changes in whistle structure of two dolphin species during interspecific associations. Ethology 116, 1065–1074.

Maynard Smith, J., 1965. The evolution of alarm calls. American Naturalist 99, 59–63.

Maze-Foley, K., Mullin, K.D., 2006. Cetaceans of the oceanic northern Gulf of Mexico: distributions, group sizes and interspecific associations. Journal of Cetacean Research and Management 8, 203–213.

McClure, H.E., 1967. The composition of mixed-species flocks in lowland and sub-montane forests of Malaya. Wilson Bulletin 79, 131–154.

McDermott, M.E., Rodewald, A.D., 2014. Conservation value of silvopastures to Neotropical migrants in Andean forest flocks. Biological Conservation 175, 140–147.

McFarland, W.N., Kotchian, N.M., 1982. Interaction between schools of fish and mysids. Behavioral Ecology and Sociobiology 11, 71–76.

McGraw, W.S., Bshary, R., 2002. Association of terrestrial mangabeys (*Cercocerbus atys*) with arboreal monkeys: experimental evidence for the effects of reduced ground predator pressure on habitat use. International Journal of Primatology 23, 311–325.

McKaye, K.R., 1985. Cichlid-catfish mutualistic defense of young in Lake Malawi, Africa. Oecologia 66, 358–363.

Menzel, F., Blüthgen, N., 2010. Parabiotic associations between tropical ants: equal partnership or parasitic exploitation? Journal of Animal Ecology 79, 71–81.

Menzel, F., Pokorny, T., Blüthgen, N., Schmitt, T., 2010. Trail-sharing among tropical ants: interspecific use of trail pheromones? Ecological Entomology 35, 495–503.

Menzel, F., Kriesell, H., Witte, V., 2014. Parabiotic ants: the costs and benefits of symbiosis. Ecological Entomology 39, 436–444.

Metcalfe, N.B., 1984. The effects of mixed-species flocking on the vigilance of shorebirds: who do they trust? Animal Behaviour 32, 986–993.

Michaelsen, T.C., Byrkjedal, I., 2002. 'Magic carpet' flight in shorebirds attacked by raptors on a migrational stopover site. Ardea 90, 167–171.

Miklósi, A., 2009. Evolutionary approach to communication between humans and dogs. Veterinary Research Communications 33, 53–59.

Mills, K.L., 1998. Multispecies seabird feeding flocks in the Galapagos Islands. Condor 100, 277–285.

Minta, S.C., Minta, K.A., Lott, D.F., 1992. Hunting associations between badgers (*Taxidea taxus*) and coyotes (*Canis latrans*). Journal of Mammalogy 73, 814–820.

Mitani, M., 1991. Niche overlap and polyspecific associations among sympatric Cercopithecids in the Campo Animal Reserve, southwestern Cameroon. Primates 32, 137–151.

Mokross, K., Ryder, T.B., Côrtes, M.C., Wolfe, J.D., Stouffer, P.C., 2014. Decay of interspecific avian flock networks along a disturbance gradient in Amazonia. Proceedings of the Royal Society of London B: Biological Sciences 281, 20132599.

Moland, E., Eagle, J.V., Jones, G.P., 2005. Ecology and evolution of mimicry in coral reef fishes. Oceanography and Marine Biology 43, 455–482.

Mönkkönen, M., Forsman, J.T., 2002. Heterospecific attraction among forest birds: a review. Ornithological Science 1, 41–51.

Mönkkönen, M., Forsman, J.T., Helle, P., 1996. Mixed-species foraging aggregations and heterospecific attraction in boreal bird communities. Oikos 77, 127–136.

Moore, F.R., Kerlinger, P., Simons, T.R., 1990. Stopover on a Gulf coast barrier island by spring trans-Gulf migrants. Wilson Bulletin 102, 487–500.

Morelli, F., Kubicka, A.M., Tryjanowski, P., Nelson, E., 2015. The vulture in the sky and the hominin on the land: three million years of human-vulture interactions. Anthrozoos 28, 449–468.

Morgan, T.C., McCleery, R.A., Moulton, M.P., Monadjem, A., 2012. Are Southern Black Flycatchers *Melaenornis pammelaina* associated with Fork-tailed Drongos *Dicrurus adsimilis*? Ostrich 83, 109–111.

Morrison, M.L., With, K.A., Timossi, I.C., Milne, K.A., 1987. Composition and temporal variation of flocks in the Sierra Nevada. Condor 89, 739–745.

Morse, D.H., 1970. Ecological aspects of some mixed-species foraging flocks of birds. Ecological Monographs 40, 119–168.

Morse, D.H., 1977. Feeding behavior and predator avoidance in heterospecific groups. BioScience 27, 332–339.

Moynihan, M., 1960. Some adaptations which help to promote gregariousness. In: Bergman, G., Donner, K.O., von Haartman, L. (Eds.), Proceedings of the XII International Ornithological Congress. Tilgmann Kirjapaino, Helsinki, pp. 523–541.

Moynihan, M., 1962. The organization and probable evolution of some mixed species flocks of neotropical birds. Smithsonian Miscellaneous Collection 143, 1–140.

Moynihan, M., 1968. Social mimicry: character convergence versus character displacement. Evolution 22, 315–331.

Mukhin, A., Chernetsov, N., Kishkinev, D., 2008. Acoustic information as a distant cue for habitat recognition by nocturnally migrating passerines during landfall. Behavioral Ecology 19, 716–723.

Munn, C.A., 1984. The Behavioral Ecology of Mixed-species Bird Flocks in Amazonian Peru (Ph.D. thesis). Princeton University, Princeton.

Munn, C.A., 1985. Permanent canopy and understorey flocks in Amazonia: species composition and population density. Ornithological Monographs 36, 683–712.

Munn, C.A., 1986. Birds that 'cry wolf'. Nature 319, 143–145.

Munn, C.A., Terborgh, J.W., 1979. Multi-species territoriality in Neotropical foraging flocks. Condor 81, 338–347.

Myers, N., Mittermeier, R.A., Mittermeier, C.G., de Fonseca, G.A.B., Kent, J., 2000. Biodiversity hotspots for conservation priorities. Nature 403, 853–858.

Newton, P.N., 1989. Associations between langur monkeys (*Presbytis entellus*) and chital deer (*Axis axis*): chance encounters or a mutualism? Ethology 83, 89–120.

Nieh, J.C., Barreto, L.S., Contrera, F.A.L., Imperatriz–Fonseca, V.L., 2004. Olfactory eavesdropping by a competitively foraging stingless bee, *Trigona spinipes*. Proceedings of the Royal Society of London B: Biological Sciences 271, 1633–1640.

Nocera, J.J., Taylor, P.D., Ratcliffe, L.M., 2008. Inspection of mob-calls as sources of predator information: response of migrant and resident birds in the Neotropics. Behavioral Ecology and Sociobiology 62, 1769–1777.

Noë, R., Bshary, R., 1997. The formation of red colobus-diana monkey associations under predation pressure from chimpanzees. Proceedings of the Royal Society of London B: Biological Sciences 264, 253–259.

Nolen, M.T., Lukas, J.R., 2009. Asymmetries in mobbing behaviour and correlated intensity during predator mobbing by nuthatches, chickadees and titmice. Animal Behaviour 77, 1137–1146.

Norris, K.S., Dohl, T.P., 1980a. Behavior of the Hawaiian spinner dolphin, *Stenella longirostris*. Fishery Bulletin 77, 821–849.

Norris, K.S., Dohl, T.P., 1980b. The structure and functions of cetacean schools. In: Herman, L.M. (Ed.), Cetacean Behavior: Mechanisms and Functions. J. H. Wiley, New York, pp. 211–261.

Nuechterlein, G.L., 1981. 'Information parasitism' in mixed colonies of western grebes and forster's terns. Animal Behaviour 29, 985–989.

Nunn, C.L., Ezenwa, V.O., Arnold, C., Koenig, W.D., 2011. Mutualism or parasitism? using a phylogenetic approach to characterize the oxpecker-ungulate relationship. Evolution 65, 1297–1304.

Obst, B.S., Hunt, G.L., 1990. Marine birds feed at gray whale mud plumes in the Bering Sea. Auk 107, 678–688.

Odling-Smee, F.J., Laland, K.N., Feldman, M.W., 2003. Niche Construction: The Neglected Process in Evolution. Princeton University Press, Princeton.

Ohtsuka, S., Inagaki, H., Onbe, T., Gushima, K., Yoon, Y.H., 1995. Direct observations of groups of mysids in shallow coastal waters of western Japan and southern Korea. Marine Ecology Progress Series 123, 33–44.

Olden, J.D., Poff, N.L., Douglas, M.R., Douglas, M.E., Fausch, K.D., 2004. Ecological and evolutionary consequences of biotic homogenization. Trends in Ecology and Evolution 19, 18–24.

Oommen, M.A., Shanker, K., 2009. Shrewd alliances: mixed foraging associations between treeshrews, greater racket-tailed drongos and sparrowhawks on Great Nicobar Island, India. Biology Letters 6, 304–307.

Orivel, J., Errard, C., Dejean, A., 1997. Ant gardens: interspecific recognition in parabiotic ant species. Behavioral Ecology and Sociobiology 40, 87–93.

Ostrand, W.D., 1999. Marbled Murrelets as initiators of feeding flocks in Prince William Sound, Alaska. Waterbirds 22, 314–318.

Otis, G.W., Locke, B., McKenzie, N.G., Cheung, D., MacLeod, E., Careless, P., Kwoon, A., 2006. Local enhancement in mud-puddling swallowtail butterflies (*Battus philenor* and *Papilio glaucus*). Journal of Insect Behavior 19, 685–698.

Overholtzer, K.L., Motta, P.J., 2000. Effects of mixed-species foraging groups on the feeding and aggression of juvenile parrotfishes. Environmental Biology of Fishes 58, 345–354.

Page, G., Whitacre, D.F., 1975. Raptor predation on wintering shorebirds. Condor 77, 73–83.

Parrish, J.K., 1989. Layering with depth in a heterospecific fish aggregation. Environmental Biology of Fishes 26, 79–85.

Partridge, L., Ashcroft, R., 1976. Mixed-species flocks of birds in hill forest in Ceylon. Condor 78, 449–453.

Patterson, J.E., Ruckstuhl, K.E., 2013. Parasite infection and host group size: a meta-analytical review. Parasitology 140, 803–813.

Paulson, D.R., 1969. Commensal feeding in grebes. Auk 86, 759.

Pays, O., Ekori, A., Fritz, H., 2014. On the advantages of mixed-species groups: Impalas adjust their vigilance when associated with larger prey herbivores. Ethology 120, 1–10.

Pennycuick, C.J., 1989. Bird Flight Performance. Oxford University Press, Oxford.

Peoples, B.K., Frimpong, E.A., 2013. Evidence of mutual benefits of nest association among freshwater cyprinids and implications for conservation. Aquatic Conservation: Marine and Freshwater Ecosystems 23, 911–923.

Pereira, P.H.C., Feitosa, J.L.L., Ferreira, B.P., 2011. Mixed-species schooling behavior and protective mimicry involving coral reef fish from the genus *Haemulon* (*Haemulidae*). Neotropical Ichthyology 9, 741–746.

Pereira, P.H.C., Feitosa, J.L.L., Chaves, L.C.T., Araújo, M.E., 2012. Reef fish foraging associations: "Nuclear-follower" behavior or an ephemeral interaction? In: 12th International Coral Reef Symposium, Cairns, pp. 1–5.

Peres, C.A., 1992. Prey-capture benefits in a mixed-species group of Amazonian tamarins, *Saguinus fuscicollis* and *Saguinus mystax*. Behavioral Ecology and Sociobiology 31, 339–347.

Peres, C.A., 1993. Anti-predation benefits in a mixed-species group of Amazonian tamarins. Folia Primatologica 61, 61–76.

Peres, C.A., 1996. Food patch structure and plant resource partitioning in interspecific associations of Amazonian tamarins. International Journal of Primatology 17, 695–723.

Périquet, S., Valeix, M., Loveridge, A.J., Madzikanda, H., Macdonald, D.W., Fritz, H., 2010. Individual vigilance of African herbivores while drinking: the role of immediate predation risk and context. Animal Behaviour 79, 665–671.

Péron, G., Crochet, P.-A., 2009. Edge effect and structure of mixed-species bird flocks in an Afrotropical lowland forest. Journal of Ornithology 150, 585–599.

Peter, C.I., Johnson, S.D., 2008. Mimics and magnets: the importance of color and ecological facilitation in floral deception. Ecology 89, 1583–1595.

Peters, M.K., Likare, S., Kraemer, M., 2008. Effects of habitat fragmentation and degradation on flocks of African ant-following birds. Ecological Applications 18, 847–858.

Pierce, N.E., Braby, M.F., Heath, A., Lohman, D.J., Mathew, J., Rand, D.B., Travassos, M.A., 2002. The ecology and evolution of ant association in the *Lycaenidae* (*Lepidoptera*). Annual Review of Entomology 47, 733–771.

Pinheiro, T., Ferrari, S.F., Lopes, M.A., 2011. Polyspecific associations between squirrel monkeys (*Saimiri sciureus*) and other primates in Eastern Amazonia. American Journal of Primatology 73, 1145–1151.

Pinto, A., Oates, J., Grutter, A., Bshary, R., 2011. Cleaner wrasses *Labroides dimidiatus* are more cooperative in the presence of an audience. Current Biology 21, 1140–1144.

Pitman, R.L., Ballance, L.T., 1992. Parkinson's Petrel distribution and foraging ecology in the eastern Pacific: aspects of an exclusive feeding relationship with dolphins. Condor 94, 825–835.

Plantan, T., Howitt, M., Kotze, A., Gaines, M., 2013. Feeding preferences of the red-billed oxpecker, *Buphagus erythrorhynchus*: a parasitic mutualist? African Journal of Ecology 51, 325–336.

Podolsky, R.D., 1990. Effects of mixed-species association on resource use by *Saimiri sciureus* and *Cebus apella*. American Journal of Primatology 21, 147–158.

Pomara, L.Y., Cooper, R.J., Petit, L.J., 2003. Mixed-species flocking and foraging behavior of four neotropical warblers in Panamanian shade coffee fields and forests. Auk 120, 1000–1012.

Pomara, L.Y., Cooper, R.J., Petit, L.J., 2007. Modeling the flocking propensity of passerine birds in two Neotropical habitats. Oecologia 153, 121–133.

Pook, A.G., Pook, G., 1982. Polyspecific association between *Saguinus fuscicollis, S. labiatus, Callimico goeldii*, and other primates in northwestern Bolivia. Folia Primatologica 38, 196–216.

Porter, L.M., 2001. Benefits of polyspecific associations for the Goeldi's monkey (*Callimico goeldii*). American Journal of Primatology 54, 143–158.

Porter, J.M., Sealy, S.G., 1981. Dynamics of seabird multispecies feeding flocks: chronology of flocking in Barkley Sound, British Columbia, in 1979. Colonial Waterbirds 4, 104–113.

Possingham, H.P., Andelman, S.J., Burgman, M.A., Medellín, R.A., Master, L.L., Keith, D.A., 2002. Limits to the use of threatened species lists. Trends in Ecology and Evolution 17, 503–507.

Poulin, R., 1999. Parasitism and shoal size in juvenile sticklebacks: conflicting selection pressures from different ectoparasites? Ethology 105, 959–968.

Poulsen, B.O., 1994. Movements of single birds and mixed-species flocks between isolated fragments of cloud forest in Ecuador. Studies on Neotropical Fauna and Environment 29, 149–160.

Poulsen, B.O., 1996. Structure, dynamics, home range and activity pattern of mixed-species bird flocks in a montane alder-dominated secondary forest in Ecuador. Journal of Tropical Ecology 12, 333–343.

Powell, G.V.N., 1985. Sociobiology and adaptive significance of interspecific foraging flocks in the Neotropics. Ornithological Monographs 36, 713–732.

Powell, L.L., Cordeiro, N.J., Stratford, J.A., 2015. Ecology and conservation of avian insectivores of the rainforest understory: a pantropical perspective. Biological Conservation 188, 1–10.

Pöysä, H., 1986a. Foraging niche shifts in multispecies dabbling duck (*Anas* spp.) feeding groups: harmful and beneficial interactions between species. Ornis Scandinavica 17, 333–346.

Pöysä, H., 1986b. Species composition and size of dabbling duck (*Anas* spp.) feeding groups: are foraging interactions important determinants? Ornis Fennica 63, 33–41.

Proctor, C.J., Broom, M., Ruxton, G.D., 2001. Modelling antipredator vigilance and flight response in group foragers when warning signals are ambiguous. Journal of Theoretical Biology 211, 409–417.

Prop, J., Quinn, J.L., 2003. Constrained by available raptor hosts and islands: density-dependent reproductive success in red-breasted geese. Oikos 102, 571–580.

Prum, R.O., 2014. Interspecific social dominance mimicry in birds. Zoological Journal of the Linnean Society 172, 910–941.

Psarakos, S., Herzing, D.L., Marten, K., 2003. Mixed-species associations between Pantropical spotted dolphins (*Stenella attenuata*) and Hawaiian spinner dolphins (*Stenella longirostris*) off Oahu, Hawaii. Aquatic Mammals 29, 390–395.

Quérouil, S., Silva, M.A., Cascão, I., Magalhães, S., Seabra, M.A., Santos, R., 2008. Why do dolphins form mixed-species associations in the Azores? Ethology 114, 1183–1194.

Quinn, J.L., Ueta, M., 2008. Protective nesting associations in birds. Ibis 150, 146–167.

Radford, A.N., Bell, M.B.V., Hollen, L.I., Ridley, A.R., 2011. Singing for your supper: sentinel calling by kleptoparasites can mitigate the cost to victims. Evolution 65, 900–906.

Ragusa-Netto, J., 1997. Evidence for the possible advantage of heterospecific social foraging in *Furnarius rufus* (*Passeriformes*: *Furnariidae*). Ararajuba 5, 233–235.

Ragusa-Netto, J., 2000. Raptors and 'Campo-Cerrado' bird mixed-flock led by *Cypsnagra hirundinacea* (*Emberizidae*: *Thraupinae*). Revista Brasileira de Biologia 60, 461–467.

Ragusa-Netto, J., 2002. Vigilance towards raptors by nuclear species in bird mixed flocks in a Brazilian savannah. Studies on Neotropical Fauna and Environment 37, 219–226.

Rainey, H.J., Zuberbühler, K., Slater, P.J.B., 2004. The responses of black-casqued hornbills to predator vocalisations and primate alarm calls. Behaviour 141, 1263–1277.

Rainho, A., Palmeirim, J.M., 2013. Prioritizing conservation areas around multispecies bat colonies using spatial modeling. Animal Conservation 16, 438–448.

Rand, A.L., 1954. Social feeding behavior of birds. Fieldiana. Zoology 36, 1–71.

Randler, C., 2004. Vigilance of Mallards in the presence of Greylag Geese. Journal of Field Ornithology 75, 404–408.

Randler, C., Förschler, M.I., 2011. Heterospecifics do not respond to subtle differences in chaffinch mobbing calls: message is encoded in number of elements. Animal Behaviour 82, 725–730.

Rasa, O.A.E., 1983. Dwarf mongoose and hornbill mutualism in the Taru Desert, Kenya. Behavioral Ecology and Sociobiology 12, 181–190.

Rautio, P., Bergvall, U.A., Tuomi, J., Kesti, K., Leimar, O., 2012. Food selection by herbivores and neighbourhood effects in the evolution of plant defences. Annales Zoologici Fennici 49, 45–57.

Regh, J.A., 2006. Seasonal variations in polyspecific associations among *Callimico goeldii*, *Saguinus labiatus*, and *S. fuscicollis* in Acre, Brazil. International Journal of Primatology 27, 1399–1428.

Rice, D.W., 1956. Dynamics of range expansion of Cattle Egrets in Florida. Auk 73, 259–266.

Richner, H., Heeb, P., 1995. Is the information center hypothesis a flop? Advances in the Study of Behavior 24, 1–44.

Ridley, A.R., Child, M.F., 2009. Specific targeting of host individuals by a kleptoparasitic bird. Behavioral Ecology and Sociobiology 63, 1119–1126.

Ridley, A.R., Raihani, N.J., 2007. Facultative response to a kleptoparasite by the cooperatively breeding pied babbler. Behavioral Ecology 18, 324–330.

Ridley, A.R., Child, M.F., Bell, M.B.V., 2007. Interspecific audience effects on the alarm-calling behaviour of a kleptoparasitic bird. Biology Letters 3, 589–591.

Ridley, A.R., Raihani, N.J., Bell, M.B.V., 2010. Experimental evidence that sentinel behaviour is affected by risk. Biology Letters 6, 445–448.

Ridley, A.R., Wiley, E.M., Thompson, A.M., 2014. The ecological benefits of interceptive eavesdropping. Functional Ecology 28, 197–205.

Ritz, D.A., 1994. Social aggregations in pelagic invertebrates. Advances in Marine Biology 30, 155–216.

Ritz, D.A., Hobday, A.J., Montgomery, J.C., Ward, A.J.W., 2011. Social aggregation in the pelagic zone with special reference to fish and invertebrates. Advances in Marine Biology 60, 161–226.

Rodgers, G.M., Kimbell, H., Morrell, L.J., 2013. Mixed-phenotype grouping: the interaction between oddity and crypsis. Oecologia 172, 59–68.

Rolando, A., Laiolo, P., Formica, M., 1997. The influence of flocking on the foraging behaviour of the chough (*Pyrrhocorax pyrrhocorax*) and the Alpine chough (*P. graculus*) coexisting in the Alps. Journal of Zoology 242, 299–308.

Roubik, D.W., 1988. Ecology and Natural History of Tropical Bees. Cambridge University Press, Cambridge.

Rubenstein, D.R., Barnett, R.J., Ridgely, R.S., Klopfer, P.H., 1977. Adaptive advantages of mixed-species feeding flocks among seed-eating finches in Costa Rica. Ibis 119, 10–21.

Ruckstuhl, K.E., 1999. To synchronise or not to synchronise: a dilemma for young bighorn males? Behaviour 136, 805–818.

Ruckstuhl, K.E., 2007. Sexual segregation in vertebrates: proximate and ultimate causes. Integrative and Comparative Biology 47, 245–257.

Ruckstuhl, K.E., Neuhaus, P., 2002. Sexual segregation in ungulates: a comparative test of three hypotheses. Biological Reviews 77, 77–96.

Russell, J.K., 1978. Effects of interspecific dominance among egrets commensally following roseate spoonbills. Auk 95, 608–610.

Ruttan, A., Lortie, C.J., 2015. A systematic review of the attractant-decoy and repellent-plant hypotheses: do plants with heterospecific neighbours escape herbivory? Journal of Plant Ecology 8, 337–346.

Rutz, C., 2012. Predator fitness increases with selectivity for odd prey. Current Biology 22, 820–824.

Ruxton, G.D., Sherratt, T.N., Speed, M.P., 2004. Avoiding Attack: The Evolutionary Ecology of Crypsis, Warning Signals and Mimicry. Oxford University Press, Oxford.

Sabino, J., Sazima, I., 1999. Association between fruit-eating fish and foraging monkeys in western Brazil. Ichthyological Exploration of Freshwaters 10, 309–312.

Safina, C., 1990. Bluefish mediation of foraging competition between Roseate and Common Terns. Ecology 71, 1804–1809.

Sakai, Y., Kohda, M., 1995. Foraging by mixed-species groups involving a small angelfish, *Centropyge ferrugatus* (*Pomacanthidae*). Japanese Journal of Ichthyology 41, 429–435.

Sasvári, L., 1992. Great tits benefit from feeding in mixed-species flocks: a field experiment. Animal Behaviour 43, 289–296.

Satischandra, S.H.K., Kudavidanage, E.P., Kotagama, S.W., Goodale, E., 2007. The benefits of joining mixed-species flocks for a sentinel nuclear species, the Greater Racket-tailed Drongo *Dicrurus paradiseus*. Forktail 23, 145–148.

Sawadogo, P.S., Namountougou, M., Toé, K.H., Rouamba, J., Maïga, H., Ouédraogo, K.R., Baldet, T., Gouagna, L.-C., Kengne, P., Simard, F., 2014. Swarming behaviour in natural populations of *Anopheles gambiae* and *An. coluzzii*: review of four years survey in rural areas of sympatry, Burkina Faso (West Africa). Acta Tropica 132, S42–S52.

Sazima, I., 2002. Juvenile snooks (*Centropomidae*) as mimics of mojarras (*Gerreidae*), with a review of aggressive mimicry in fishes. Environmental Biology of Fishes 65, 37–45.

Sazima, I., 2013. Five instances of bird mimicry suggested for Neotropical birds: a brief reappraisal. Revista Brasileira de Ornitologia 18, 328–335.

Sazima, I., Sazima, C., 2010. Cleaner birds: an overview for the Neotropics. Biota Neotropica 10, 196–203.

Sazima, C., Krajewski, J.P., Bonaldo, R.M., Sazima, I., 2007. Nuclear-follower foraging associations of reef fishes and other animals at an oceanic archipelago. Environmental Biology of Fishes 80, 351–361.

Scheel, D., 1993. Watching for lions in the grass: the usefulness of scanning and its effects during hunts. Animal Behaviour 46, 695–704.

Schlupp, I., Marler, C., Ryan, M.J., 1994. Benefit to male sailfin mollies of mating with heterospecific females. Science 263 (5145), 373–374.

Schlupp, I., Ryan, M.J., 1996. Mixed-species shoals and the maintenance of a sexual–asexual mating system in mollies. Animal Behaviour 52, 885–890.

Schmitt, M.H., Stears, K., Wilmers, C.C., Shrader, A.M., 2014. Determining the relative importance of dilution and detection for zebra foraging in mixed-species herds. Animal Behaviour 96, 151–158.

Schmitt, M.H., Stears, K., Shrader, A.M., 2016. Zebra reduce predation risk in mixed-species herds by eavesdropping on cues from giraffe. Behavioral Ecology 27, 1073–1077.

Schneider, D., 1982. Fronts and seabird aggregations in the southeastern Bering Sea. Marine Ecology Progress Series 10, 101–103.

Schreffler, L., Leiser, J.K., Master, T.L., 2010. Costs and benefits of foraging alone or in mixed-species aggregations for Forster's Terns. Wilson Journal of Ornithology 122, 95–101.

Schummer, M.L., Petrie, S.A., Bailey, R.C., 2008. Interactions between macroinvertebrate abundance and habitat use by diving ducks during winter on northeastern Lake Ontario. Journal of Great Lakes Research 34, 54–71.

Scott, M.D., Cattanach, K.L., 1998. Diel patterns in aggregations of pelagic dolphin and tuna in the eastern Pacific. Marine Mammal Science 14, 401–428.

Scott, M.D., Chivers, S.J., Olson, R.J., Fiedler, P.C., Holland, K., 2012. Pelagic predator associations: tuna and dolphins in the eastern tropical Pacific Ocean. Marine Ecology Progress Series 458, 283–302.

Sealey, S.G., 1973. Interspecific feeding assemblages of marine birds off British Columbia. Auk 90, 796–802.

Seddon, N., Tobias, J.A., Eaton, M., Odeen, A., 2010. Human vision can provide a valid proxy for avian perception of sexual dichromatism. Auk 127, 283–292.

Şekercioğlu, C.H., 2006. Increasing awareness of avian ecological function. Trends in Ecology and Evolution 21, 464–471.

Semeniuk, C.A.D., Dill, L.M., 2006. Anti-predator benefits of mixed-species groups of cowtail stingrays (*Pastinachus sephen*) and whiprays (*Himantura uarnak*) at rest. Ethology 112, 33–43.

Seppänen, J.-T., Forsman, J.T., 2007. Interspecific social learning: novel preference can be acquired from a competing species. Current Biology 17, 1248–1252.

Seppänen, J.-T., Forsman, J.T., Mönkkönnen, M., Thomson, R.L., 2007. Social information use is a process across time, space and ecology, reaching heterospecifics. Ecology 88, 1622–1633.

Seppanen, J.T., Forsman, J.T., Monkkonen, M., Krams, I., Salmi, T., 2011. New behavioural trait adopted or rejected by observing heterospecific tutor fitness. Proceedings of the Royal Society of London B: Biological Sciences 278, 1736–1741.

Seyfarth, R.M., Cheney, D.L., Marler, P., 1980. Vervet monkey alarm calls: semantic communication in a free-ranging primate. Animal Behaviour 28, 1070–1094.

Sharpe, L.L., Joustra, A.S., Cherry, M.I., 2010. The presence of an avian co-forager reduces vigilance in a cooperative mammal. Biology Letters 6, 475–477.

Shelden, K.E.W., Baldridge, A., Withrow, D.E., 1995. Observations of Risso dolphins, *Grampus giseus* with gray whales, *Eschrichtius robustus*. Marine Mammal Science 11, 231–240.

Shultz, S., Faurie, C., Noë, R., 2003. Behavioural responses of Diana monkeys to male long-distance calls: changes in ranging, association patterns and activity. Behavioral Ecology and Sociobiology 53, 238–245.

Sidhu, S., Raman, T.R.S., Goodale, E., 2010. Effects of plantations and home-gardens on tropical forest bird communities and mixed-species bird flocks in the southern Western Ghats. Journal of the Bombay Natural History Society 107, 91–108.

Sieving, K.E., Contreras, T.A., Maute, K.L., 2004. Heterospecific facilitation of forest boundary crossing by mobbing understory birds in north-central Florida. Auk 121, 738–751.

Sigel, B.J., Sherry, T.W., Young, B.E., 2006. Avian community response to lowland tropical rainforest isolation: 40 years of change at La Selva Biological Station, Costa Rica. Conservation Biology 20, 111–121.

Silvano, R.A.M., 2001. Feeding habits and interspecific feeding associations of *Caranx latus* (*Carangidae*) in a subtropical reef. Environmental Biology of Fishes 60, 465–470.

Silverman, E.D., Veit, R.R., 2001. Associations among Antarctic seabirds in mixed species feeding flocks. Ibis 143, 51–62.

Silverman, E.D., Kot, M., Thompson, E., 2001. Testing a simple stochastic model for the dynamics of waterfowl aggregations. Oecologia 128, 608–617.

Silverman, E.D., Veit, R.R., Nevitt, G.A., 2004. Nearest neighbors as foraging cues: information transfer in a patchy environment. Marine Ecology Progress Series 277, 25–35.

Simberloff, D., 1998. Flagships, umbrella, and keystones: is single-species management passé in the landscape era? Biological Conservation 83, 247–257.

Sinclair, A.R.E., 1985. Does interspecific competition or predation shape the African ungulate community? Journal of Animal Ecology 54, 899–918.

Sinu, P.A., 2011. Avian pest control in tea plantations of sub-Himalayan plains of Northeast India: mixed-species foraging flock matters. Biological Control 58, 362–366.

Slaa, E.J., Wassenberg, J., Biesmeijer, J.C., 2003. The use of field-based social information in eusocial foragers: local enhancement among nestmates and heterospecifics in stingless bees. Ecological Entomology 28, 369–379.

Smith, N.G., 1968. The advantage of being parasitized. Nature 219, 690–694.

Smith, J., 1995. Foraging sociability of nesting wading birds (Ciconiiformes) at Lake Okeechobee, Florida. Wilson Bulletin 107, 437–457.

Smith, A.C., Buchanan-Smith, H.M., Surridge, A.K., Mundy, N.I., 2003. Leaders of progressions in wild mixed-species troops of saddleback (*Saguinus fuscicollis*) and mustached tamarins (*S. mystax*), with emphasis on color vision and sex. American Journal of Primatology 61, 145–157.

Sonerud, G.A., Smedshaug, C.A., Bråthen, Ø., 2001. Ignorant hooded crows follow knowledgeable roost-mates to food: support for the information centre hypothesis. Proceedings of the Royal Society of London B: Biological Sciences 268, 827–831.

Spottiswoode, C.N., Begg, K.S., Begg, C.M., 2016. Reciprocal signaling in honeyguide-human mutualism. Science 353, 387–389.

Sridhar, H., Sankar, K., 2008. Effects of habitat degradation on mixed-species bird flocks in Indian rain forests. Journal of Tropical Ecology 24, 135–147.

Sridhar, H., Shanker, K., 2014a. Importance of intraspecifically gregarious species in a tropical bird community. Oecologia 176, 763–770.

Sridhar, H., Shanker, K., 2014b. Using intra-flock association patterns to understand why birds participate in mixed-species foraging flocks in terrestrial habitats. Behavioral Ecology and Sociobiology 68, 185–196.

Sridhar, H., Beauchamp, G., Shanker, K., 2009. Why do birds participate in mixed-species foraging flocks? A large-scale synthesis. Animal Behaviour 78, 337–347.

Sridhar, H., Srinivasan, U., Askins, R.A., Canales-Delgadillo, J.C., Chen, C.-C., Ewert, D.N., Gale, G.A., Goodale, E., Gram, W.K., Hart, P.J., Hobson, K.A., Hutto, R.L., Kotagama, S.W., Knowlton, J.L., Lee, T.M., Munn, C.A., Nimnuan, S., Nizam, B.Z., Péron, G., Robin, V.V., Rodewald, A.D., Rodewald, P.G., Thomson, R.L., Trivedi, P., Van Wilgenburg, S.L., Shanker, K., 2012. Positive relationships between association strength and phenotypic similarity characterize the assembly of mixed-species bird flocks worldwide. American Naturalist 180, 777–790.

Sridhar, H., Jordán, F., Shanker, K., 2013. Species importance in a heterospecific foraging association network. Oikos 122, 1325–1334.

Srinivasan, U., Quader, S., 2012. To eat and not be eaten: modelling resources and safety in multi-species animal groups. PLOS ONE 7, e42071.

Srinivasan, U., Raza, R.H., Quader, S., 2010. The nuclear question: rethinking species importance in multi-species animal groups. Journal of Animal Ecology 79, 948–954.

Srinivasan, U., Raza, R.H., Quader, S., 2012. Patterns of species participation across multiple mixed-species flock types in a tropical forest in northeastern India. Journal of Natural History 46, 2749–2762.

Srygley, R.B., Penz, C.M., 1999. Lekking in neotropical owl butterflies, *Caligo illioneus* and *C. oileus* (*Lepidoptera*: *Brassolinae*). Journal of Insect Behavior 12, 81–103.

Stachowicz, J.J., Hay, M.E., 1999. Mutualism and coral persistence: the role of herbivore resistance to algal chemical defense. Ecology 80, 2085–2101.

Stamps, J.A., 1988. Conspecific attraction and aggregation in territorial species. American Naturalist 329–347.

Stawarczyk, T., 1984. Aggression and its suppression in mixed-species wader flocks. Ornis Scandinavica 15, 23–37.

Stensland, E., Angerbjörn, A., Berggren, P., 2003. Mixed species groups in mammals. Mammalogy Reviews 33, 205–223.

Stinson, C.H., 1980. Flocking and predator avoidance: models of flocking and observations on the spatial dispersion of foraging wintering shorebirds (Charadrii). Oikos 34, 35–43.

Stojan-Dolar, M., Heymann, E.W., 2010. Vigilance of mustached tamarins in single-species and mixed-species groups – the influence of group composition. Behavioral Ecology and Sociobiology 64, 325–335.

Stouffer, P.C., Bierregaard, R.O., 1995. Use of Amazonian forest fragments by understory insectivorous birds. Ecology 76, 2429–2445.

Stouffer, P.C., Bierregaard, R.O., 1996. Forest fragmentation and seasonal patterns of hummingbird abundance in Amazonian Brazil. Ararajuba 4, 9–14.

Stout, J.C., Goulson, D., 2001. The use of conspecific and interspecific scent marks by foraging bumblebees and honeybees. Animal Behaviour 62, 183–189.

Strand, S., 1988. Following behavior: interspecific foraging associations among Gulf of California reef fishes. Copeia 351–357.

Struhsaker, T.T., 1981. Polyspecific associations among tropical rain-forest primates. Zeitschrift für Tierpsychologie 57, 268–304.

Sullivan, K.A., 1984. Information exploitation by downy woodpeckers in mixed-species flocks. Behaviour 91, 294–311.

Sumpter, D.J., 2010. Collective Animal Behavior. Princeton University Press, Princeton.

Suzuki, T.N., 2012. Long-distance calling by the willow tit, *Poecile montanus*, facilitates formation of mixed-species foraging flocks. Ethology 118, 10–16.

Suzuki, T.N., 2016. Referential calls coordinate multi-species mobbing in a forest bird community. Journal of Ethology 34, 79–84.

Swartz, M.B., 2001. Bivouac checking, a novel behavior distinguishing obligate from opportunistic species of army-ant-following birds. Condor 103, 629–633.

Teelen, S., 2007. Influence of chimpanzee predation on associations between red colobus and red-tailed monkeys at Ngogo, Kibale National Park, Uganda. International Journal of Primatology 28, 593–606.

Tellería, J.L., Virgós, E., Carbonell, R., Pérez-Tris, J., Santos, T., 2001. Behavioural responses to changing landscapes: flock structure and anti-predator strategies of tits wintering in fragmented forests. Oikos 95, 253–264.

Templeton, C.N., Greene, E., 2007. Nuthatches eavesdrop on variations in heterospecific chickadee mobbing alarm calls. Proceedings of the National Academy of Sciences of the United States of America 104, 5479–5482.

Templeton, C.N., Greene, E., Davis, K., 2005. Allometry of alarm calls: black-capped chickadees encode information about predator size. Science 308, 1934–1937.

Terborgh, J., 1983. Five New World Primates: A Study in Comparative Ecology. Princeton University Press, Princeton.

Terborgh, J., 1990. Mixed flocks and polyspecific associations: costs and benefits of mixed groups to birds and monkeys. American Journal of Primatology 21, 87–100.

Tews, J., Brose, U., Grimm, V., Tielborger, K., Wichmann, M.C., Schwager, M., Jeltsch, F., 2004. Animal species diversity driven by habitat heterogeneity/diversity: the importance of keystone structures. Journal of Biogeography 31, 79–92.

Thiollay, J.-M., 1992. Influence of selective logging on bird species diversity in a Guianan rain forest. Conservation Biology 6, 47–60.

Thiollay, J.-M., 1999. Frequency of mixed-species flocking in tropical forest birds and correlates of predation risk: an intertropical comparison. Journal of Avian Biology 30, 282–294.

Thiollay, J.-M., 2003. Comparative foraging behavior between solitary and flocking insectivores in a Neotropical forest: does vulnerability matter? Ornitologia Neotropical 14, 47–65.

Thiollay, J.-M., Jullien, M., 1998. Flocking behaviour of foraging birds in a neotropical rain forest and the antipredator defence hypothesis. Ibis 140, 382–394.

Thompson, D.B.A., Barnard, C.J., 1983. Anti-predator responses in mixed-species associations of lapwings, golden plovers and black-headed gulls. Animal Behaviour 31, 585–593.

Thompson, D.B.A., Thompson, M.L.P., 1985. Early warning and mixed-species association: the 'Plover's Page' revisited. Ibis 127, 559–562.

Thorpe, W.H., 1956. Learning and Instinct in Animals. Methuen, London.

Tisovec, K.C., Cassano, C.R., Boubli, J.P., Pardini, R., 2014. Mixed-species groups of marmosets and tamarins across a gradient of agroforestry intensification. Biotropica 46, 248–255.

Tosh, C.R., Jackson, A.L., Ruxton, G.D., 2007. Individuals from different-looking animal species may group together to confuse shared predators: simulations with artificial neural networks. Proceedings of the Royal Society of London B: Biological Sciences 274, 827–832.

Tubelis, D.P., 2007. Mixed-species flocks of birds in the Cerrado, South America: a review. Ornitologia Neotropical 18, 75–97.

Tubelis, D.P., Cowling, A., Donnelley, C., 2006. Role of mixed-species flocks in the use of adjacent savannas by forest birds in the central Cerrado, Brazil. Austral Ecology 31, 38–45.

Turcotte, Y., Desrochers, A., 2002. Playbacks of mobbing calls of black-capped chickadees help estimate the abundance of forest birds in winter. Journal of Field Ornithology 73, 303–307.

Turner, G.F., Pitcher, T.J., 1986. Attack abatement: a model for group protection by combined avoidance and dilution. American Naturalist 128, 228–240.

Tuttle, M.D., Ryan, M.J., 1981. Bat predation and the evolution of frog vocalizations in the Neotropics. Science 214, 677–678.

Tylianakis, J.M., Laliberté, E., Nielsen, A., Bascompte, J., 2009. Conservation of species interaction networks. Biological Conservation 143, 2270–2279.

Ueda, H., Kuwahara, A., Tanaka, M., Azeta, M., 1983. Underwater observations on copepod swarms in temperate and subtropical waters. Marine Ecology Progress Series 11, 165–171.

Uetz, G.W., 1989. The "ricochet effect" and prey capture in colonial spiders. Oecologia 81, 154–159.

Ulfstrand, S., 1975. Bird flocks in relation to vegetation diversification in a south Swedish coniferous plantation during winter. Oikos 26, 65–73.

Underwood, N., Inouye, B.D., Hamback, P.A., 2014. A conceptual framework for associational effects: when do neighbors matter and how would we know? Quarterly Review of Biology 89, 1–19.

Vail, A.L., Manica, A., Bshary, R., 2013. Referential gestures in fish collaborative hunting. Nature Communications 4, 1765.

Valiente-Banuet, A., Aizen, M.A., Alcántara, J.M., Arroyo, J., Cocucci, A., Galetti, M., García, M.B., García, D., Gómez, J.M., Jordano, P., 2015. Beyond species loss: the extinction of ecological interactions in a changing world. Functional Ecology 29, 299–307.

Van Houtan, K.S., Pimm, S.L., Bierregaard Jr., R.O., Lovejoy, T.E., Stouffer, P.C., 2006. Local extinctions in flocking birds in Amazonian forest fragments. Evolutionary Ecology Research 8, 129–148.

Vaughn, R.L., Shelton, D.E., Timm, L.L., Watson, L.A., Würsig, B., 2007. Dusky dolphin (*Lagenorhynchus obscurus*) feeding tactics and multi-species associations. New Zealand Journal of Marine and Freshwater Research 41, 391–400.

Veena, T., Lokesha, R., 1993. Association of drongos with myna flocks: are drongos benefitted? Journal of Biosciences 18, 111–119.

Veit, R.R., Silverman, E.D., Everson, I., 1993. Aggregation patterns of pelagic predators and their principal prey, Antarctic krill, near South Georgia. Journal of Animal Ecology 62, 551–564.

Videler, J.J., Wardle, C.S., 1991. Fish swimming stride by stride: speed limits and endurance. Reviews in Fish Biology and Fisheries 1, 23–40.

Vieth, W., Curio, E., Ernst, U., 1980. The adaptive significance of avian mobbing. III. cultural transmission of enemy recognition in blackbirds: cross-species tutoring and properties of learning. Animal Behaviour 28, 1217–1229.

Viscido, S.V., Shrestha, S., 2015. Using quantitative methods of determining group membership to draw biological conclusions. Animal Behaviour 104, 145–154.

Vitousek, M.N., Adelman, J.S., Gregory, N.C., St Clair, J.J.H., 2007. Heterospecific alarm call recognition in a non-vocal reptile. Biology Letters 3, 632–634.

Vuilleumier, F., 1967. Mixed species flocks in Patagonian forests, with remarks on interspecies flock formation. Condor 69, 400–404.

Waite, T.A., Grubb, T.C., 1988. Copying of foraging location in mixed-species flocks of temperate-deciduous woodland birds: an experimental study. Condor 90, 132–140.

Wallace, A.R., 1875. Contributions to the Theory of Natural Selection: A Series of Essays. MacMillan and Co., London.

Wang, Z.W., Wen, P., Qu, Y.F., Dong, S.H., Li, J.J., Tan, K., Nieh, J.C., 2016. Bees eavesdrop upon informative and persistent signal compounds in alarm pheromones. Scientific Reports 6, 25693.

Ward, A.J.W., Krause, J., 2001. Body length assortative sorting in the European minnow, *Phoxinus phoxinus*. Animal Behaviour 62, 617–621.

Ward, P., Zahavi, A., 1973. The importance of certain assemblages of birds as "information-centres" for food-finding. Ibis 115, 517–534.

Ward, A.J.W., Axford, S., Krause, J., 2002. Mixed-species shoaling in fish: the sensory mechanisms and costs of shoal choice. Behavioral Ecology and Sociobiology 52, 182–187.

Ward, A.J.W., Axford, S., Krause, J., 2003. Cross-species familiarity in shoaling fishes. Proceedings of the Royal Society of London B: Biological Sciences 270, 1157–1161.

Ward, A.J.W., Hart, P.J.B., Krause, J., 2004. The effects of habitat- and diet-based cues on association preferences in three-spined sticklebacks. Behavioral Ecology 15, 925–929.

Waser, P.M., 1984. "Chance" and mixed-species associations. Behavioral Ecology and Sociobiology 15, 197–202.

Waser, P.M., Case, T.J., 1981. Monkeys and matrices: on the coexistence of "omnivorous" forest primates. Oecologia 49, 102–108.

Waterman, J.M., Roth, J.D., 2007. Interspecific associations of Cape ground squirrels with two mongoose species: benefit or cost? Behavioral Ecology and Sociobiology 61, 1675–1683.

Webster, M.M., Ward, A.J.W., Hart, P.J.B., 2008. Shoal and prey patch choice by co-occurring fishes and prawns: inter-taxa use of socially transmitted cues. Proceedings of the Royal Society of London B: Biological Sciences 275, 203–208.

Weeks, P., 2000. Red-billed oxpeckers: vampires or tickbirds? Behavioral Ecology 11, 154–160.

Weimerskirch, H., Martin, J., Clerquin, Y., Alexandre, P., Jiraskova, S., 2001. Energy saving in flight formation - Pelicans flying in a 'V' can glide for extended periods using the other birds' air streams. Nature 413, 697–698.

Weimerskirch, H., Bertrand, S., Silva, J., Marques, J.C., Goya, E., 2010. Use of social information in seabirds: compass rafts indicate the heading of food patches. PLoS ONE 5, e9928.

Wertheim, B., van Baalen, E.-J.A., Dicke, M., Vet, L.E.M., 2005. Pheromone-mediated aggregation in nonsocial arthropods: an evolutionary ecological perspective. Annual Review of Entomology 50, 321–346.

Westcott, P.W., 1969. Relationships among three species of jays wintering in southeastern Arizona. Condor 71, 353–359.

Westrip, J.R.S., Bell, M.B.V., 2015. Breaking down the species boundaries: selective pressures behind interspecific communication in vertebrates. Ethology 121, 725–732.

Wey, T., Blumstein, D.T., Shen, W., Jordan, F., 2008. Social network analysis of animal behaviour: a promising tool for the study of sociality. Animal Behaviour 75, 333–344.

Wheatcroft, D., Price, T.D., 2013. Learning and signal copying facilitate communication among bird species. Proceedings of the Royal Society of London B: Biological Sciences 280, 20123070.

Whiting, M.J., Greeff, J.M., 1999. Use of heterospecific cues by the lizard *Platysaurus broadleyi* for food location. Behavioral Ecology and Sociobiology 45, 420–423.

Wiley, R.H., 1971. Cooperative roles in mixed flocks of antwrens (*Formicariidae*). Auk 88, 881–892.

Willard, D.E., 1977. The feeding ecology and behavior of five species of herons in southeastern New Jersey. Condor 79, 462–470.

Willis, E.O., 1963. Is the zone-tailed hawk a mimic of the Turkey vulture? Condor 65, 313–317.

Willis, E.O., 1972. The behavior of spotted antbirds. Ornithological Monographs 10, 1–162.

Willis, E.O., 1973a. Do birds flock in Hawaii, a land without predators? California Birds 3, 1–8.

Willis, E.O., 1973b. Local distribution of mixed-species flocks in Puerto Rico. Wilson Bulletin 85, 75–77.

Willis, E.O., 1976. Similarity of a tyrant-flycatcher and a silky-flycatcher: not all character convergence is competitive mimicry. Condor 74, 553.

Willis, E.O., Oniki, Y., 1978. Birds and army ants. Annual Review of Ecology and Systematics 9, 243–263.

Willis, P.M., 2013. Why do animals hybridize? Acta Ethologica 16, 127–134.

Willson, S.K., 2004. Obligate army-ant-following birds: a study of ecology, spatial movement patterns, and behavior in Amazonian Peru. Ornithological Monographs 55, 1–67.

Wilson, E.O., 1975. Sociobiology: The New Synthesis. Belknap Press, Cambridge, MA.

Wilson, S.E., Allen, J.A., Anderson, K.P., 1990. Fast movement of densely aggregated prey increases the strength of anti-apostatic selection by wild birds. Biological Journal of the Linnean Society 41, 375–380.

Windfelder, T.L., 2001. Interspecific communication in mixed-species groups of tamarins: evidence from playback experiments. Animal Behaviour 61, 1193–1201.

Wing, L., 1946. Species associations in winter flocks. Auk 63, 507–510.

Winterbottom, J.M., 1943. On woodland bird parties in northern Rhodesia. Ibis 84, 437–442.

Winterbottom, J.M., 1949. Mixed bird parties in the tropics, with special reference to Northern Rhodesia. Auk 66, 258–263.

Wittmann, K.J., 1977. Modification of association and swarming in North Adriatic Mysidacea in relation to habitat and interacting species. In: Keegan, B.F., O'Ceidigh, P., Boaden, P.J.S. (Eds.), Biology of Benthic Organisms. Pergamon Press, Oxford, pp. 605–612.

Wolf, N.G., 1985. Odd fish abandon mixed-species groups when threatened. Behavioral Ecology and Sociobiology 17, 47–52.

Wolters, S., Zuberbühler, K., 2003. Mixed-species associations of Diana and Campbell's monkeys: the costs and benefits of a forest phenomenon. Behaviour 140, 371–385.

Wood, C., Sullivan, B., Iliff, M., Fink, D., Kelling, S., 2011. eBird: engaging birds in science and conservation. PLoS Biology 9, e1001220.

Wrege, P.H., Wikelski, M., Mandel, J.T., Rassweiler, T., Couzin, I.D., 2005. Antbirds parasitize foraging army ants. Ecology 86, 555–559.

Yao, I., 2014. Costs and constraints in aphid-ant mutualism. Ecological Research 29, 383–391.

Yodzis, P., 1978. Competition for Space and the Structure of Ecological Communities. Springer-Verlag, Berlin.

Zamon, J.E., 2003. Mixed species aggregations feeding upon herring and sandlance schools in a nearshore archipelago depend on flooding tidal currents. Marine Ecology Progress Series 261, 243–255.

Zamora, R., Hodar, J.A., Gómez, J.M., 1992. Dartford Warblers follow Stonechats while foraging. Ornis Scandinavica 23, 167–174.

Zhang, Q., Han, R., Huang, Z., Zou, F., 2013. Linking vegetation structure and bird organization: response of mixed-species bird flocks to forest succession in subtropical China. Biodiversity and Conservation 22, 1965–1989.

Zuberbühler, K., 2000. Interspecies semantic communication in two forest primates. Proceedings of the Royal Society of London B: Biological Sciences 267, 713–718.

Zuberbühler, K., Noë, R., Seyfarth, R.M., 1997. Diana monkey long-distance calls: messages for conspecifics and predators. Animal Behaviour 53, 589–604.

Index

Printed in the United States
By Bookmasters